SpringerBriefs in Oceanography

For further volumes:
http://www.springer.com/series/11754

Arthur N. Popper • Anthony D. Hawkins
Richard R. Fay • David A. Mann
Soraya Bartol • Thomas J. Carlson
Sheryl Coombs • William T. Ellison
Roger L. Gentry • Michele B. Halvorsen
Svein Løkkeborg • Peter H. Rogers
Brandon L. Southall • David G. Zeddies
William N. Tavolga

ASA S3/SC1.4 TR-2014

Sound Exposure Guidelines for Fishes and Sea Turtles:

A Technical Report prepared by ANSI-Accredited Standards Committee S3/SC1 and registered with ANSI

Registered with ANSI on 20 April 2014

ASA Press

Springer

Arthur N. Popper
Department of Biology, University of Maryland
College Park, MD 20742, USA
apopper@umd.edu

Anthony D. Hawkins
Loughine Ltd, Aberdeen, AB12 5YT, UK
a.hawkins@btconnect.com

Richard R. Fay
Marine Biological Laboratory
Woods Hole Oceanographic Institution
Woods Hole, MA 02540, USA
rfay44@aol.com

David A. Mann
Loggerhead Instruments, Sarasota, FL 34233, USA
dmann@loggerheadinstruments.com

Soraya Bartol
Biology Department, Virginia Wesleyan College
Norfolk, VA 23502, USA
sbartol@vwc.edu

Thomas J. Carlson
ProBioSound LLC, Holmes Beach, FL 34217, USA
tj3carlson@gmail.com

Sheryl Coombs
Department of Biological Sciences
Bowling Green State University
Bowling Green, OH 43403, USA
scoombs@bgnet.bgsu.edu

William T. Ellison
Marine Acoustics, Inc., Middletown, RI 03801, USA
WEMAI@aol.com

Roger L. Gentry
ProScience Consulting, LLC
Dickerson, MD 20842, USA
Roger.gentry@comcast.net

Michele B. Halvorsen
CSA Ocean Sciences, Stuart, FL 34997, USA
mhalvy@gmail.com

Svein Løkkeborg
Institute of Marine Research, Bergen 05005, Norway
svein.loekkeborg@imr.no

Peter H. Rogers
School of Mechanical Engineering
Georgia Institute of Technology
Atlanta, GA 30332, USA
peter.rogers@me.gatech.edu

Brandon L. Southall
Southall Environmental Associates Inc.
Aptos, CA 95003, USA
brandon.southall@sea-inc.net

David G. Zeddies
JASCO Applied Sciences
Silver Spring, MD 20902, USA
david.zeddies@jasco.com

William N. Tavolga
Sensory Biology and Behavior
Mote Marine Laboratory, Sarasota, FL 34231, USA
Tavolga@aol.com

ISSN 2196-1212 ISSN 2196-1220 (electronic)
ISBN 978-3-319-06658-5 ISBN 978-3-319-06659-2 (eBook)
DOI 10.1007/978-3-319-06659-2
Springer Cham Heidelberg New York Dordrecht London

Library of Congress Control Number: 2014940282

Printed on acid-free paper

Springer is part of Springer Science+Business Media (www.springer.com)

Acoustical Society of America

The mission of the **Acoustical Society of America** (www.acousticalsociety.org) is to increase and diffuse the knowledge of acoustics and promote its practical applications. The ASA is recognized as the world's premier international scientific society in acoustics, and counts among its more than 7,000 members professionals in the fields of bioacoustics, engineering, architecture, speech, music, oceanography, signal processing, sound and vibration, and noise control.

Since its first meeting in 1929, The Acoustical Society of America has enjoyed a healthy growth in membership and in stature. The present membership of approximately 7,000 includes leaders in acoustics in the United States of America and other countries. The Society has attracted members from various fields related to sound including engineering, physics, oceanography, life sciences, noise and noise control, architectural acoustics; psychological and physiological acoustics; applied acoustics; music and musical instruments; speech communication; ultrasonics, radiation, and scattering; mechanical vibrations and shock; underwater sound; aeroacoustics; macrosonics; acoustical signal processing; bioacoustics; and many more topics.

To assure adequate attention to these separate fields and to new ones that may develop, the Society establishes technical committees and technical groups charged with keeping abreast of developments and needs of the membership in their specialized fields. This diversity and the opportunity it provides for interchange of knowledge and points of view has become one of the strengths of the Society.

The Society's publishing program has historically included the *Journal of the Acoustical Society of America*, the magazine *Acoustics Today* (www.acousticstoday.org), a newsletter, and various books authored by its members across the many topical areas of acoustics. In addition, ASA members are involved in the development of acoustical standards concerned with terminology, measurement procedures, and criteria for determining the effects of noise and vibration.

Abstract

This Technical Report presents the outcome of a Working Group that was established to determine broadly applicable sound exposure guidelines for fishes and sea turtles. After consideration of the diversity of fish and sea turtles, guidelines were developed for broad groups of animals, defined by the way they detect sound. Different sound sources were considered in terms of their acoustic characteristics and appropriate metrics defined for measurement of the received levels. The resultant sound exposure guidelines are presented in a set of tables. In some cases numerical guidelines are provided, expressed in appropriate metrics. When there were insufficient data to support numerical values, the *relative* likelihood of effects occurring was evaluated, although the *actual* likelihood of effects depends on the received level. These sound exposure guidelines, which are based on the best scientific information at the time of writing, should be treated as interim. The expectation is that with more research, the guidelines can be refined and more cells in the tables completed. Recommendations are put forward defining the research requirements of highest priority for extending these interim exposure guidelines.

Foreword

[*This Foreword is for information only, and is not a part of the Technical Report ASA S3/SC1.4 TR-2014 Sound Exposure Guidelines for Fishes and Sea Turtles.*]

This Technical Report comprises a part of a group of definitions, standards, and specifications for use in animal bioacoustics. It was developed and approved by Accredited Standards Committee S3/SC 1 Animal Bioacoustics, under its approved operating procedures. Those procedures have been accredited by the American National Standards Institute (ANSI). The Scope of Accredited Standards Committee S3/SC 1 is as follows:

> Standards, specifications, methods of measurement and test, instrumentation, and terminology in the field of psychological and physiological acoustics, including aspects of general acoustics which pertain to biological safety, tolerance, and comfort of non-human animals, including both risk to individual animals and to the long-term viability of populations. Animals to be covered may potentially include commercially grown food animals; animals harvested for food in the wild; pets; laboratory animals; exotic species in zoos, oceanaria or aquariums; or free-ranging wild animals.

Publication of this Technical Report that has been registered with ANSI has been approved by the ANSI-Accredited Standards Committee S3/SC 1, Animal Bioacoustics. This document is registered as a Technical Report according to the Procedures for the Registration of Technical Reports with ANSI. This document is not an American National Standard and the material contained herein is not normative in nature. Comments on the content of this document should be sent to Standards Secretariat of the Acoustical Society of America, 1305 Walt Whitman Rd., Ste. 300 Melville, New York 11747-4300. Telephone: 631-390-0215; FAX: 631-923-2875; E-mail: asastds@aip.org.

As required by ASC S3/SC 1's operating procedures, this Technical Report will be subjected to periodic review by ASC S3/SC 1 at least every five years. However, it is possible that the committee will revise or withdraw it at any time. Users should check to confirm if a newer version is available.

The rationale for publishing this technical report is detailed in the Introduction.

At the time this Technical Report was submitted to Accredited Standards Committee S3/SC 1, Animal Bioacoustics, for approval the membership was as follows:

D.K. Delaney, *Chair*
M.C. Hastings, *Vice Chair*
S.B. Blaeser, *Secretary*

Acoustical Society of America	M.C. Hastings
American Academy of Otolaryngology, Head and Neck Surgery	R.A. Dobie
American Speech-Language-Hearing Association	B.M. Ryals N. DiSarno (Alt.)
G.R.A.S. Sound & Vibration	J. Soendergaard B. Schustrich (Alt.)
International Association of Geophysical Contractors	S. Tsoflias
National Oceanographic and Atmospheric Administration	J. Gedamke A. Scholik-Schlomer (Alt.)
National Park Service	K. Fristrup M. McKenna (Alt.)
Ocean Conservation Research	M. Stocker
SpeakDolphin.com	J. Kassewitz S. Kuczaj (Alt.)
U.S. Army Aeromedical Research Lab	W.A. Ahroon
U.S. Army Construction Engineering Research Laboratory	D.K. Delaney M.J. White (Alt.)
U.S. Navy Marine Mammal Program	J.J. Finneran D.S. Houser (Alt.)
University of Cincinnati Animal Audiology Clinic/Bioacoustics Lab	P.M. Scheifele D.K. Brown (Alt.)

Working Group S3/SC 1/WG 2, Effects of Sound on Fish and Turtles, which assisted Accredited Standards Committee S3/SC 1, Animal Bioacoustics, in the development of this Technical Report, had the following membership.

R.R. Fay, Co-Chair
A.N. Popper, Co-Chair

S. Bartol
T.J. Carlson
S. Coombs
W.T. Ellison
R.L. Gentry
M.B. Halvorsen
A.D. Hawkins
S. Løkkeborg
D.A. Mann
P.H. Rogers
B.L. Southall
W.N. Tavolga
D.G. Zeddies

Keywords

Acoustics, Active Sonar, Anthropogenic, Airgun, Air-gun, Audiogram, Auditory Scene Analysis, Barotrauma, Behavior, Criteria, Ear, Eggs, Damage, Directional Hearing, Explosions, Explosives, Fitness, Frequency Weighting, Hearing, Injury, Intensity, Larvae, Lateral Line, Masking, Noise, Particle Motion, Peak, Pile Driving, Pressure, PTS, RMS, Scene Analysis, Seismic Airgun, Ship, Sonar, Sound Exposure Level, Swim Bladder, Threshold, Threshold Shift, TTS

Acknowledgements

The need for this panel was identified by Roger Gentry, then at NOAA in the U.S. Department of Commerce. He and Brandon L. Southall, also then at NOAA, identified the original panel members and allocated NOAA funds for its first three meetings. Subsequently, support for the meetings was provided by the National Science Foundation, the Office of Naval Research, and the Joint Industry Program (JIP) of the Oil and Gas Producers Association. Additional funds and support were provided by the Acoustical Society of America (ASA). We are grateful to all of these agencies for their support. Most of our meetings took place at the Mote Marine Laboratory in Sarasota, Florida. We want to thank the staff at Mote for their hospitality and their efforts to make the meetings comfortable and productive. We are grateful to members of the ANSI-Accredited Standards Committee S3/SC 1, Animal Bioacoustics, which is sponsored by the Acoustical Society of America, for their thorough and very useful review of this MS. Finally, we thank Dr. Peggy Edds-Walton for editing the final MS and Susan Blaeser of ASA for additional thoughtful suggestions.

Acknowledgements

Contents

1 Introduction 1
1.1 Background 1
1.2 Background Literature 2
1.3 Terminology 3

2 Aquatic Organisms of Concern 5
2.1 Fishes 5
2.2 Sea Turtles 6
2.3 Eggs and larvae 6

3 Hearing – A General Overview 7
3.1 Fish Hearing: Sensitivity to Sound 7
3.2 Lateral Line System 10
3.3 Turtle Hearing 11
3.4 Masking 12
3.5 Auditory Scene Analysis 13
3.6 Directional Hearing 13

4 Classification of Fishes and Sea Turtles with Respect to Sound Exposure Risk 15
4.1 Fishes 15
4.2 Sea Turtles 16
4.3 Larval Fish 16
4.4 Categories 16

5 Effects of Sound Exposure 17
5.1 Death and Injury 17
5.2 Effects on Hearing 18
5.3 Effects on Behavior 19
5.4 Population-Level Effects on Fitness and Survival 20

6 The Nature of Man-Made Sound 23
6.1 Metrics 24
6.1.1 Sound Pressure 24
6.1.2 Particle Motion 24
6.1.3 Peak Levels 25
6.1.4 Sound Intensity 25
6.1.5 Impulse 26
6.1.6 Sound Exposure Level 26
6.1.7 Cumulative Energy (Cumulative Sound Exposure Level) 26
6.1.8 Frequency Weighting 27
6.2 Sources 28
6.2.1 Explosions 28
6.2.2 Seismic Airguns 29
6.2.3 Pile Driving 30
6.2.4 Active Sonar 31
6.2.5 Continuous Sound Sources 31

7 Sound Exposure Guidelines 33
7.1 Current Guidelines for Injury 33
7.2 Current Guidelines for Behavior 35
7.3 Practical Approach to Recommending Sound Exposure Guidelines 36
7.4 Explanation of the Analyses 37
7.5 Guidelines 37
7.5.1 Source: Explosions 37
7.5.2 Source: Pile Driving 41
7.5.3 Source: Seismic Airguns 44
7.5.4 Source: Low- and Mid-Frequency Naval Sonar 47
7.5.5 Source: Shipping and Other Continuous Noises 49

8 Research Recommendations 53
8.1 General Comments 53
8.1.1 The Importance of Providing an Appropriate Acoustic Environment for Experiments 53
8.1.2 The Difficulties in Examining Behavior 54
8.1.3 The Diversity of Fishes and Sea Turtles 55
8.2 Research Requirements of Highest Priority 55
8.2.1 Fishes 55
8.2.2 Eggs and Larvae 57
8.2.3 Sea Turtles 57

9 Summary and Conclusions 59

References 61

Chapter 1
Introduction

1.1 Background

In many countries, legislation requires the assessment of sound-producing activities that may have an impact on animals in the aquatic environment (TNO 2011; Johnson 2012; Lewandowski et al. 2012; Tasker 2012, 2015; Dekeling et al. 2015; Gedamke et al. 2015). There is also often a requirement to prepare environmental assessments or statements that can lead to mitigation measures and/or restrictions for proposed activities. Because few scientific data are available regarding the effects of sound, particularly for fishes and sea turtles, assessment procedures and subsequent regulatory and mitigation measures are often severely limited in their relevance and efficacy. This creates uncertainty among all stakeholders as to how sound-producing exploration and operations should proceed.

In 1998 the U.S. National Oceanic and Atmospheric Administration (NOAA) convened an international panel of experts to prepare noise exposure guidelines for marine mammals. NOAA's intent was to provide its regulatory staff with the means of issuing permits on underwater noise production based on a set of organized principles and facts instead of on a case-by-case basis. The aim was to reduce regulatory uncertainty for all stakeholders by replacing precaution with scientific facts. The panel's initial guidelines appeared in a seminal paper (Southall et al. 2007).

In 2004, NOAA convened a similar panel to develop noise exposure criteria for fishes and turtles with the same goals in mind. It included three members of the marine mammal panel (WTE, RLG, and BLS) for continuity. When NOAA's support for this effort ended in 2006, the panel was organized as a Working Group (WG) under the ANSI-Accredited Standards Committee S3/SC 1, Animal Bioacoustics, which is sponsored by the Acoustical Society of America.

In addition, this Working Group, through its own efforts, obtained external funding for the project (see Acknowledgements). The Working Group met eight times between 2004 and 2010. It gathered and reviewed papers from both the peer-review and grey literature that presented data on the exposure of fish and sea turtles to

A.N. Popper et al., *ASA S3/SC1.4 TR-2014 Sound Exposure Guidelines for Fishes and Sea Turtles*, SpringerBriefs in Oceanography, DOI 10.1007/978-3-319-06659-2_1,

various sound sources. Subsequent smaller meetings were held from 2011 to 2013 to review and finalize this manuscript. During that time, findings were updated and the revised document was continuously reviewed by the whole Working Group. The guidelines presented here are primarily based on data in the peer-reviewed literature up until mid-2013.

These Guidelines represents the Working Group's consensus efforts to establish broadly applicable sound exposure guidelines for fishes and sea turtles across the complete range of taxa and sound types, and to consider many impacts, rather than just injury. The Working Group adopts some of the general approaches used to set sound exposure guidelines for marine mammals (Southall et al. 2007). However, several factors make the present Guidelines differ in format and conclusions from Southall et al. (2007). These factors include:

1) There are more than 32,000 species of fish compared to about 130 species of marine mammals;
2) Fishes are much more diverse anatomically, physiologically, ecologically, and behaviorally than are marine mammals;
3) Most fishes respond to the particle motion component of sound waves whereas marine mammals do not;
4) Relatively few papers link exposure to effects in fishes; and
5) While there are few species of sea turtles, so little is known about their hearing and the role of sound in their lives that it is very difficult to establish guidelines for these species.

These guidelines are not intended to be a complete review of the literature. Rather, the material cited is limited to those publications that provide background to help explain how and why the guidelines were selected. Readers interested in more complete reviews are directed to references mentioned in Chap. 1.2. As in Southall et al. (2007), this report does not consider the commercial, societal, and practical considerations of the conclusions reached. Instead, these Guidelines serve as the first step in setting guidelines that may lead to the establishment of exposure standards for fishes and sea turtles.

1.2 Background Literature

These Guidelines are written for scientists, regulators, environmentalists and others who may not have a detailed knowledge of hearing or bioacoustics of fishes and sea turtles. For readers without a background in bioacoustics, several web sites provide lay-level information, especially the Discovery of Sound in the Sea site (www.dosits.org) and/or the Aquatic Acoustic Archive (www.aquaticacousticarchive.com/). More extensive reviews are cited in Hawkins et al. (2008), Popper and Hawkins (2012, 2015), and the report of a recent BOEM (Bureau of Ocean Energy Management) workshop (Normandeau Associates 2012a, b). Additional background material can be found in Webb et al. (2008), Au and Hastings (2008), and Coombs et al. (2014).

A background on general effects of noise is found in Le Prell et al. (2012), while a background in underwater acoustics can be found in Urick (1983) and Au and Hastings (2008). A less technical account on underwater sound can be found in Rogers and Cox (1988) and at www.dosits.org.

1.3 Terminology

In this review the term "fishes" refers to more than one species, whereas "fish" refers to one or more members of a single species. A number of different expressions are used to describe sounds produced by humans. For this report, we generally refer to "man-made" sound, meaning any sound with a source caused by human activity.

The term "*noise*" is most often used to describe undesired sound, or sound that interferes with detection of any sound that is of interest. However, noise has also been used to describe background levels of sound in the sea, including the naturally occurring sounds generated by distributed biological sources, weather events or physical phenomena like ice ridging, some of which cannot be assigned to individual sources. In this report the term "*sound*" is used both to refer to identifiable man-made sources such as individual ships or oil and gas platforms, or to distant man-made sources that cannot be located or identified. Where others have used the terms "ambient noise" or "background noise" to distinguish naturally occurring sounds from distributed sources, that usage will be respected.

Chapter 2
Aquatic Organisms of Concern

2.1 Fishes

There are over 32,000 extant fish species (Helfman et al. 2009; www.fishbase.org), more than all other vertebrate groups combined. Fishes show extensive variability in their behavior, ecology, and physiology. Moreover, fishes vary in their abilities to detect and utilize sounds, and very likely also vary in their potential susceptibility to damage by sound. The taxonomy and general biology of fishes is set out in Helfman et al. (2009).

Many factors are likely to be important in the effects of sound exposure and their longer-term consequences for fitness and survival, but one of the most important of these is the presence or absence of a gas bladder in the body (see also Chap. 3.1). Gas bladders, and their anatomical location within the body, make fish more susceptible to pressure-mediated (sound pressure and barotrauma[1]) injury to the ears and general body tissues than species lacking gas bladders (Stephenson et al. 2010; Halvorsen et al. 2011; Carlson 2012). The presence of a gas bladder is also likely to increase the ability of many species of fish to detect sounds over a broader frequency range and at greater distances from the source than fishes without such structures, thereby increasing the range from the source over which man-made sound sources have the potential to exert influence (Chap. 3.1).

[1] As discussed in more detail in Chap. 5.1, barotrauma is tissue injury that results from rapid pressure changes (e.g., forced change in depth, explosions and intense sound) (e.g., Stephenson et al. 2010; Halvorsen et al. 2011, 2012a).

A.N. Popper et al., *ASA S3/SC1.4 TR-2014 Sound Exposure Guidelines for Fishes and Sea Turtles*, SpringerBriefs in Oceanography, DOI 10.1007/978-3-319-06659-2_2,

2.2 Sea Turtles

There are seven extant species of sea turtle: green (*Chelonia mydas*), hawksbill (*Eretmochelys imbricate*), Kemp's Ridley (*Lepidochelys kempii*), loggerhead (*Caretta caretta*), olive Ridley (*Lepidochelys olivacea*), flatback sea turtle (*Natator depressus*), and leatherback (*Dermochelys coriacea*). All of these generally share a similar body form, although shell morphology is different in leatherback turtles compared to the hard-shelled species.

There are few data on hearing abilities of sea turtles (Chap. 3.3), their uses of sound, and their vulnerability to sound exposure. It has thus been necessary to extrapolate from other animal groups. Though there has been some discussion of using data from marine mammals to predict turtle responses (e.g., Finneran and Jenkins 2012), it is the view of the WG that, while still unsatisfactory, data from fishes provide a better analogy at this time (see also Chap. 3.3). The rationale is that the hearing range for turtles much more approximates that of fishes than of any marine mammal, and the functioning of the basilar papilla in the turtle ear is dissimilar to the functioning of the cochlea in mammals.

2.3 Eggs and larvae

We have separated out fish eggs and larvae for special consideration because of their vulnerability, reduced mobility, and small size. Very few peer-reviewed papers discuss the responses of eggs and larvae to man-made sound.

Chapter 3
Hearing – A General Overview

3.1 Fish Hearing: Sensitivity to Sound[1]

All fishes have ears to detect sound and convey sensitivity to gravity and to linear and angular acceleration (for more extensive reviews of fish hearing see Fay, 1988; Fay and Megela Simmons 1999; Popper et al. 2003; Popper and Schilt 2008; Fay and Edds-Walton 2008; Sand and Bleckmann 2008). The auditory portions of the ears are the otolithic organs (saccule, lagena, and utricle), with the saccule being the primary auditory organ in many, but not all, species. Each otolithic organ consists of a dense calcareous mass contacting a sensory epithelium. Otolithic organs of all fishes respond to particle motion of the surrounding fluid. As sound impinges upon the fish, the otoliths' greater density and rigidity causes them to move at a lower amplitude and different phase than the surrounding tissue. Many fishes are also able to detect sound pressure via the gas bladder or other gas-filled structures that re-radiate energy, in the form of particle motion, to the otolithic organs (e.g., van Bergeijk 1967; Chapman and Sand 1974; Rogers and Zeddies 2008). Particle motion is a vector quantity with both magnitude and direction, and represents the oscillatory displacement (m), velocity (m/s), or acceleration (m/s^2) of fluid particles in a sound field. In contrast, pressure is a scalar quantity with magnitude only, and is a form of stress that is measured in terms of force/unit area (N/m^2 or Pa). (See Chap. 6.1 for further discussion of sound).

Lying close to each otolith is a sensory epithelium that has embedded in it numerous receptors: the sensory hair cells (reviewed in Popper et al. 2003). Each sensory hair cell has a bundle of stereovilli (often called stereocilia) projecting toward, and often contacting, the otolith. Relative motion between the otolith and the epithelium results in a deflection of the cilia, thereby activating the hair cells. All fishes detect particle motion in this way.

[1] The literature often refers to "hearing specialists" and "hearing generalists," but this differentiation has been shown to be inappropriate (Popper and Fay 2011) and will not be used herein.

A.N. Popper et al., *ASA S3/SC1.4 TR-2014 Sound Exposure Guidelines for Fishes and Sea Turtles,* SpringerBriefs in Oceanography, DOI 10.1007/978-3-319-06659-2_3,

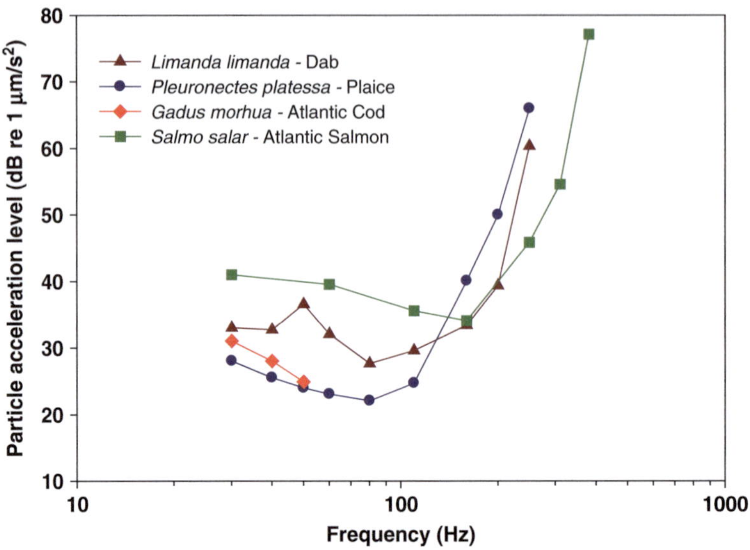

Fig. 3.1 Particle motion behavioral audiograms for four species that are particle motion-sensitive, or where sensitivity to particle motion is dominant at the frequencies plotted. Atlantic salmon (*Salmo salar*, Hawkins and Johnstone 1978); plaice (*Pleuronectes platessa*, Chapman and Sand 1974); dab (*Limanda limanda*, Chapman and Sand 1974); Atlantic cod (*Gadus morhua*, Chapman and Hawkins 1973)

Figure 3.1 provides audiograms that illustrate the sensitivity (lowest sound level detected) and the frequency range of hearing of several particle motion-sensitive species including Atlantic salmon (*Salmo salar*), plaice (*Pleuronectes platessa*), dab (*Limanda limanda*), and, at the lowest frequencies, Atlantic cod (*Gadus morhua*).

The adaptations that provide fish with a sensitivity to sound pressure are gas-filled structures near the ear and/or extensions of the swim bladder that functionally affect the ear. The enclosed gas changes volume in response to fluctuating sound pressure, generating particle motion. In fishes where the swim bladder is near the ear (or connected to it mechanically as in the Otophysi), the particle motion radiated from the bladder is sufficiently large to cause the sensory epithelium to move relative to the otolith. Fishes with these adaptations generally have lower sound pressure thresholds and wider frequency ranges of hearing (Fig. 3.2) than do the purely particle motion-sensitive species (Fig. 3.1).

The ability to detect sound pressure in addition to particle motion serves to increase hearing sensitivity and broaden the hearing bandwidth (e.g., Sand and Enger 1973a, b; Sand and Hawkins 1973, 1974; Fletcher and Crawford 2001). Examples of species that detect sound pressure as well as particle motion include Atlantic cod (Chapman and Hawkins 1973; Sand and Enger 1973a, b), European eel (*Anguilla anguilla*) (Jerkø et al. 1989), and various species of damselfish (Family

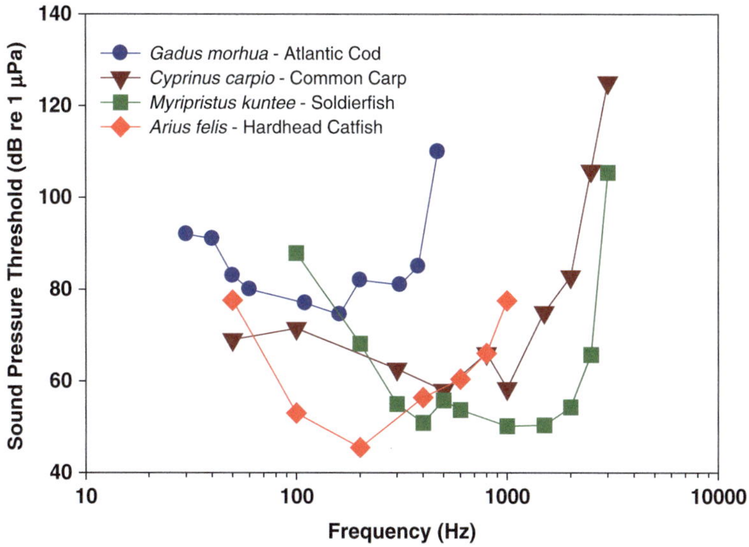

Fig. 3.2 Sound pressure behavioral audiograms for four species that are sound pressure-sensitive in the frequency regions plotted. Common carp (*Cyprinus carpio*, Popper 1972); Atlantic cod (*Gadus morhua*, Chapman and Hawkins 1973); soldierfish (*Myripristus kuntee,* Coombs and Popper 1979); hardhead catfish (*Arius felis*, Popper and Tavolga 1981)

Pomacentridae) (Myrberg and Spires 1980). Some fishes that have a swim bladder only detect particle motion (e.g., Atlantic salmon) (Hawkins and Johnstone 1978). However, the relative importance of particle motion vs. sound pressure in the hearing capabilities of such species is likely to vary, and is at least in part related to the distance and connection between the anterior portion of the swim bladder and the inner ear (Popper and Fay 2011).

Hearing range and sensitivity varies considerably among species (Figs. 3.1 and 3.2). Some species with a swim bladder are sound pressure-sensitive at higher frequencies (the Atlantic cod), while others having a swim bladder are not (Atlantic salmon). At least three species of herring-like fishes (but not the Atlantic herring *Clupea harengus* itself) detect sounds above 20 kHz (ultrasound). But for most species, sensitivity to sound occurs from below 100 Hz to several hundred hertz, or several thousand hertz in a few species (Mann et al. 1997, 2001). Predicting approximate hearing sensitivity based on the anatomy of the ear and swim bladder is not always possible due to the extraordinary variations found in the anatomy of the ears and swim bladders in different species.

Behavioral audiograms have been published for only a few species of fish and there are concerns about the usefulness of many of these. These concerns arise for two reasons. First, many of these audiograms were obtained under poorly monitored acoustic conditions and it is difficult to determine whether the fish were responding to sound pressure or particle motion, as Rogers et al. (2015) point out. Secondly, many audiograms were determined in conditions where background noise was not measured. Noise can result in the audiograms being masked (see Chap. 3.4) so that the full hearing sensitivity of the animal cannot be determined. In fact, even in Figure 3.2, the thresholds reported at some frequencies for the cod may be limited by masking from ambient sea noise.

A related issue is that many contemporary hearing studies measure sensitivity in terms of auditory evoked potentials (AEP) that are recorded from the ear or brainstem (reviewed in Ladich and Fay 2013; Sisneros et al. 2015). However, AEP recordings may not fully reflect the hearing capabilities of animals (e.g., AEPs do not include signal processing by the brain), and they often do not mirror audiograms obtained by behavioral experiments. For these reasons, AEP data should be used with caution in predicting the actual behavioral hearing sensitivity of the animal and/or the degree of effect a particular man-made noise might have on fishes (Sisneros et al. 2015).

3.2 Lateral Line System

The lateral line system consists of hundreds to thousands of flow sensors distributed all over the head and body of bony and cartilaginous fish, as well as aquatic and semi-aquatic amphibians (Cahn 1967; Coombs et al. 1989, 2014; Montgomery and Coombs 1996; Coombs and Montgomery 1999; Montgomery et al. 2014). The most conspicuous part of the lateral line system is a canal, the openings to which are often seen along the length of the fish's body. This and other canals on the head contain groups of hair cell sensors (neuromasts) evenly distributed along the length of the canal. In addition to the canals, most fishes also have free neuromasts distributed on the surface of the skin.

Each lateral line neuromast has many displacement-sensitive sensory hair cells that are essentially the same as those found in the ears of all vertebrates. Receptors of the lateral line system respond to the relative motion between the body surface and the surrounding water. This relative motion only takes place very close to sound sources where there is a steep gradient of sound pressure and particle motion (Denton and Gray 1982, 1993; Kalmijn 1988). As a consequence, the operational range of the lateral line is usually restricted to no more than one or two body lengths away from the source.

The lateral line plays an important role in predator avoidance maneuvers, prey detection, courtship and spawning, orienting to the direction of water flow (rheotaxis), station holding in currents, and spatial imaging and exploration in the absence of vision (reviewed in Dijkgraaf 1963; Bleckmann 1993; Montgomery and Coombs 1996; Coombs and Montgomery 1999; Sand and Bleckmann 2008; Montgomery et al. 2014).

There have been no demonstrations to date of damage to lateral line systems as a result of exposure to intense man-made sounds or other signals (Hastings et al. 1996), although it is conceivable that damage may occur. Some man-made sound sources, such as moving bodies (e.g., ships) and stationary bodies in a flow field (e.g., construction piles in a water current) can generate damming phenomena and hydrodynamic wakes that may function as both signals and maskers for the lateral line system.

3.3 Turtle Hearing

Data on hearing by sea turtles is very limited. Electrophysiological studies on hearing have been conducted on juvenile green sea turtles (Ridgway et al. 1969; Bartol and Ketten 2006; Piniak et al. 2012), juvenile Kemp's Ridleys (Bartol and Ketten 2006), and on juvenile loggerheads (Bartol et al. 1999; Lavender et al. 2012). Ridgway et al. (1969) obtained an AEP audiogram to aerial and vibrational stimuli that extended from below 100 Hz to 2000 Hz with the lowest threshold at 400 Hz. Other studies using AEPs found similar low-frequency responses to vibrations delivered to the tympanum (the external ear on the surface of the head) for the loggerhead sea turtle (Bartol et al. 1999) and to underwater sound stimuli for the loggerhead, Kemp's Ridley, and green sea turtles (Bartol and Ketten 2006; Bartol and Bartol 2011; Lavender et al. 2012).

Christiansen-Dalsgaard et al. (2012) used physiological methods to measure hearing in an amphibious fresh water turtle, the red-ear slider (*Trachemys scripta elegans*). They found that the tympanic ear was specialized for underwater sound-pressure hearing with greatest sensitivity at frequencies between 500 to 600 Hz (see also Willis 2015).

Martin et al. (2012) measured underwater thresholds in the loggerhead sea turtle (*Caretta caretta*) by both behavioral and AEP methods. Behavioral sensitivity showed the lowest thresholds between 100 and 400 Hz, with thresholds at about 100 dB re 1 μPa. AEP measurements on the same individual were up to 8 dB higher; however, both techniques showed a similar frequency response and a high-frequency loss of sensitivity above 400 Hz of about 37 dB per octave (Fig. 3.3).

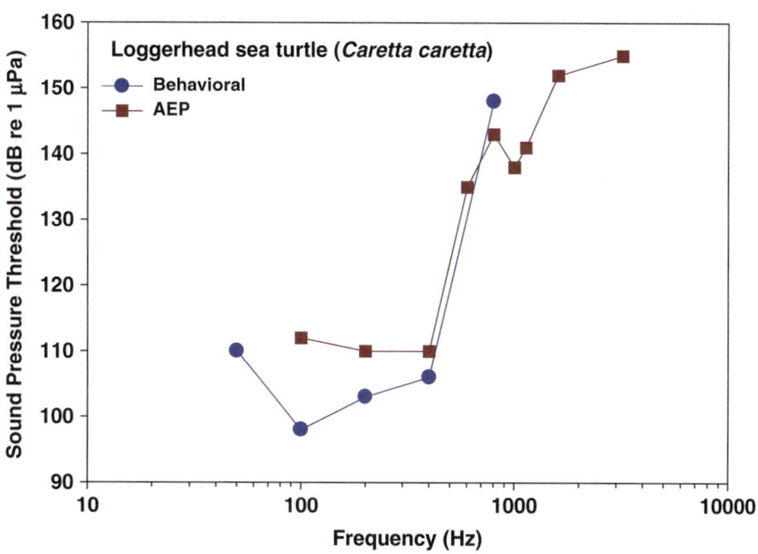

Fig. 3.3 Behavioral and auditory evoked potential thresholds for the Loggerhead sea turtle (Martin et al. 2012)

Other preliminary measures of hearing in sea turtles indicated that the hearing range was 50 - 1200 Hz (Lavender et al. 2012).

Morphological examinations of green and loggerhead sea turtles (Ridgway et al. 1969; Wever 1978; Lenhardt et al. 1985) describe the sea turtle as having a typical reptilian ear with a few underwater modifications, supporting the proposal that fish hearing, rather than mammalian hearing, is the better model to use for sea turtles until there are much more data. The tympanum, on the surface of the head, is backed by a thick layer of subtympanal fat, a feature that distinguishes sea turtles from both terrestrial and semi-aquatic turtles. The middle ear cavity is connected to the throat by a Eustachian tube (Wever 1978; Lenhardt et al. 1985). The ossicular bones of the middle ear connect the tympanum to the inner ear. The bones consist of two elements, the columella and the extracolumella. For semi-aquatic turtles, the columella is the main pathway for sound input to the inner ear, and when the columella is clipped while leaving the tympanum intact, the animal displays a substantial decrease in hearing sensitivity (Wever and Vernon 1956)

The auditory sense organ within the inner ear of the sea turtle is the basilar papilla, and also possibly the saccule (Wever and Vernon 1956). The basilar papilla is positioned opposite the round window and lies within the pathway of fluid displacement that results from motion of the columella in response to vibration of the tympanic membrane.

3.4 Masking

Masking is the reduction in the detectability of a given sound (signal) as a result of the simultaneous occurrence of another sound (noise). One of the most critical issues with regard to behavioral effects of sound on fishes and sea turtles is whether man-made sound interferes with, or masks, the ability of the animal to detect and respond to biologically relevant sounds. In effect, masking raises the threshold for detection by an animal. Masking may occur where a masking noise exceeds the absolute hearing thresholds of an animal, and is likely to occur for most fishes and sea turtles at some locations and times as a result of the varying level of ambient noise that occurs in all aquatic environments (Fay 1974). In water, this ambient noise may come from such sources as waves breaking on the shore, undersea earthquakes, precipitation, water moving across the substrate, animal vocalizations, etc. For example, masking of sound detection by ambient noise in the sea has been demonstrated for the Atlantic cod (Hawkins and Chapman 1975). In the natural world, all types of sounds may be both signals of interest or interfering noise maskers, depending on context. Masking of sound detection has been termed energetic masking if it results from interaction between signal and masker at the periphery of the auditory system. Informational masking, refers to the interpretation of information from masker and signal by the brain (e.g., Durlach et al. 2003), resulting in uncertainties about the characteristics of a complex signal as a result of the presence of a masker. The masking of various biologically significant sounds by man-made sounds is probably more extensive than would be predicted from energetic masking alone (Fay 2010).

Data on hearing for all vertebrates tested to date, including fishes, show that the degree of masking is related both to the level of the masking noise and the frequencies it contains. In fishes, pure tone sounds are masked most readily by noise at the same and immediately adjacent frequencies, falling within a critical band (as shown by Tavolga 1967, 1974; Hawkins and Chapman 1975). There is a complete lack of data on masking of biologically important signals in fish and sea turtles by man-made sounds with differing characteristics.

3.5 Auditory Scene Analysis

Auditory scene analysis (Bregman 1990) is an important aspect of auditory reception in humans and animals. The term refers to the perceptual scene (but not necessarily a scene with spatial dimensions) made up of all the animate and inanimate sources producing or scattering sounds that are detectable by an animal. Auditory scene analysis involves the animal perceiving its acoustical surroundings, or soundscape, as a collection of different sources. It is assumed to exist for all vertebrate animals (e.g., Fay 1998, 2000, 2008; Fay and Popper 2000). Organisms that hear tend to segregate many of the different, simultaneous sound sources that impinge on them to form this perceptual scene. It is also assumed that animals perceiving these individual acoustic sources have a greater chance of behaving appropriately with respect to them, and thus a greater chance of survival.

By performing auditory scene analysis, an animal can potentially perceive the soundscape as a collection of sources, thereby giving the animal critical information for environmental assessment, orientation, and navigation (Slabbekoorn and Bouton 2008; Clark et al. 2009; Fay 2009; Slabbekoorn et al. 2010). The critical point is that man-made sounds may disrupt scene analysis as well as mask signals of importance to an animal (Fay 2010).

3.6 Directional Hearing

Sound source localization, or directional hearing, is one of the most important aspects of hearing for all animals, including fishes (Fay 2000; Sand and Bleckmann 2008) and, presumably, turtles. It enables an animal to move toward a food source or a mate, or away from a potential predator (Fay 2005). Directional hearing in fishes is based on the detection of particle motion (reviewed by Schuijf 1975; Sand and Bleckmann 2008). The degree of masking of a signal by any noise source also depends on the spatial location of the noise source relative to the signal source (Chapman 1973; Chapman and Johnstone 1974; Hawkins and Sand 1977). When noise and signal sources are spatially separated, masking can be reduced by 7 to 15 dB.

Chapter 4
Classification of Fishes and Sea Turtles with Respect to Sound Exposure Risk

4.1 Fishes

To develop exposure guidelines it is first necessary to place fishes in categories depending on how they might be affected by sounds. We propose categories based on the presence or absence of a swim bladder and on the potential for that swim bladder to improve the hearing sensitivity and range of hearing (Chap. 3.1). Before discussing the specific classification, however, additional background on hearing mechanisms in fishes is required.

Most bony fishes have a swim bladder, used for buoyancy control, located in the abdominal cavity that is used for buoyancy control. The swim bladder may also be used for hearing, sound production, and/or respiration. Species with a swim bladder and other gas chambers have a greater potential to suffer from physiological trauma (barotrauma) than those without gas chambers. Sudden pressure changes, whether from hydrostatic pressure or sound pressure, can cause rapid motion of the walls of these cavities, particularly from impulsive sounds. These movements can result in damage to nearby tissues such as the kidney and gonads. Additional effects result from the motion of small bubbles that may occur in the circulatory system as a result of trauma. Fish species that lack a gas-filled cavity, including jawless fishes, elasmobranchs (sharks, skates, and rays – Casper et al. 2012a), some flatfishes, some gobies, and some tuna and other pelagic and deep-sea species, are not as vulnerable to trauma from extreme sound pressure changes as fish with a gas-filled space. This difference has been demonstrated by comparing the effects of pile driving sounds on fishes with and without a swim bladder (Halvorsen et al. 2012c).

Fish species lacking a gas-filled cavity primarily detect particle motion and do not detect sound pressure (e.g., Chapman and Sand 1974 in Fig. 3.1; Hawkins and MacLennan 1976; Casper and Mann 2007a, b, 2009). Some species have a functional physical connection between the swim bladder, or some other gas chamber, and the inner ear (Popper and Coombs 1982; Braun and Grande, 2008). The hearing abilities of these animals depend much more upon sound pressure, although such fishes would also be expected to have particle motion sensitivity comparable to

A.N. Popper et al., *ASA S3/SC1.4 TR-2014 Sound Exposure Guidelines for Fishes and Sea Turtles*, SpringerBriefs in Oceanography, DOI 10.1007/978-3-319-06659-2_4,

other fishes. Moreover, since these species hear better than others, they are more likely to have their hearing affected by lower levels of masking, potentially resulting in greater behavioral effects.

4.2 Sea Turtles

As discussed in Chap. 3.3, the ear of sea turtles appears to be adapted to detect sound in water. The retention of air in the middle ear of these sea turtles suggests that they are able to detect sound pressure.

4.3 Larval Fish

While few data are available on larval fishes, those species studied appear to have hearing frequency ranges similar to those of adults (Higgs et al. 2002; Egner and Mann 2005; Zeddies and Fay 2005; Wright et al. 2011), and similar acoustic startle thresholds (Zeddies and Fay 2005). Swim bladders may develop during the larval stage and may render larvae susceptible to pressure-related injuries (e.g., barotrauma). Current concern over the effects of sound upon eggs, and especially for larvae containing gas bubbles, is focused on barotrauma (see below) rather than hearing. Only very few publications consider the effects of sound or vibration on fish eggs and larvae (Banner and Hyatt 1973; Kostyuchenko 1973; Bennett et al. 1994; Govoni et al. 2003, 2008; Jørgensen et al. 2005; Bolle et al. 2012).

4.4 Categories

Based on the above discussion, animals have been grouped into the following categories for analyzing the effects of sounds upon them:

- *Fishes with no swim bladder or other gas chamber* (e.g., dab and other flatfish). These species are less susceptible to barotrauma and only detect particle motion, not sound pressure. However, some barotrauma may result from exposure to sound pressure.
- *Fishes with swim bladders in which hearing does not involve the swim bladder or other gas volume* (e.g., Atlantic salmon). These species are susceptible to barotrauma although hearing only involves particle motion, not sound pressure.
- *Fishes in which hearing involves a swim bladder or other gas volume* (e.g., Atlantic cod, herring and relatives, Otophysi). These species are susceptible to barotrauma and detect sound pressure as well as particle motion.
- *Sea turtles*
- *Fish eggs and larvae*

Chapter 5
Effects of Sound Exposure

Sound, at higher intensities, may have a diverse range of effects on the animal. These may include death, hearing impairment, damage to anatomical structures, and changes in physiology, neural function, behavior, and development (Knight and Swaddle 2011).

5.1 Death and Injury

Death and injury can result from exposure to very high amplitude sounds (Carlson and Johnson 2010). In addition, the effects of changes in pressure (barotrauma) must also be considered, especially for impulsive sounds.

Barotrauma is tissue injury that results from rapid pressure changes (e.g., forced change in depth, explosions, and intense sound) (e.g., Stephenson et al. 2010; Halvorsen et al. 2011, 2012a). Rapid changes in pressure can cause blood gases to come out of solution. Rapid pressure changes can also cause gas volumes (i.e., swim bladders) to expand and contract rapidly, thereby damaging surrounding tissues and organs, and sometimes causing rupture of the swim bladder itself.

Injury to fish from barotrauma can be quite variable, both in cause and effect, depending upon of the pattern of pressure changes and the physiological state of the exposed fish (Stephenson et al. 2010; Halvorsen et al. 2011, 2012a, c). Sudden changes in pressure are more likely to result in damage than are gradual changes.

Barotrauma endpoints include lethal injury through immediate mortality or delayed mortality (McKinstry et al. 2007) and a number of injuries with varying severity from which full recovery is possible (e.g., Halvorsen et al. 2011, 2012a, c; Brown et al. 2012; Casper et al. 2012b, 2013a). Injuries that are potentially recoverable, such as fin hematomas, capillary dilation, and loss of sensory hair cells may still lead to death if they decrease fitness and the animal is subject to predation or disease. Mortality as a result of reduced fitness that leads to predation or disease is

A.N. Popper et al., *ASA S3/SC1.4 TR-2014 Sound Exposure Guidelines for Fishes and Sea Turtles*, SpringerBriefs in Oceanography, DOI 10.1007/978-3-319-06659-2_5,

classified as indirect mortality, whereas death as a result of injuries is classified as direct mortality (Halvorsen et al. 2011, 2012a).

Extreme levels of particle motion arising from various impulsive sources may also have the potential to injure tissues although this has yet to be demonstrated for any source.

5.2 Effects on Hearing

Hearing loss can be permanent or temporary. Permanent loss of hearing may be a consequence of the death of the sensory hair cells in the ear, damage to the innervating auditory nerve fibers (Liberman 2015) or damage to other tissues in the auditory pathway such as the swim bladder.

Temporary threshold shift (TTS) is a temporary reduction in hearing sensitivity caused by exposure to intense sound. TTS has been demonstrated in some fishes, and its extent is of variable duration and magnitude. TTS results from temporary changes in sensory hair cells of the inner ear and/or damage to auditory nerves innervating the ear (Smith et al. 2006; Liberman 2015). However, sensory hair cells are constantly added in fishes (e.g., Corwin 1981, 1983; Popper and Hoxter 1984; Lombarte and Popper 1994) and also replaced when damaged (Lombarte et al. 1993; Smith et al. 2006; Schuck and Smith 2009), unlike in the auditory receptors of mammals. When sound-induced hair cell death occurs in fishes, its effects may be mitigated over time by the addition of new hair cells (Smith et al. 2006, 2011; Smith 2012, 2015).

After termination of a sound that causes TTS, normal hearing ability returns over a period that is variable, depending on many factors, including the intensity and duration of sound exposure (e.g., Popper and Clarke 1976; Scholik and Yan 2001,2002a, b; Amoser and Ladich, 2003; Smith et al. 2004a, b, 2006, 2011; Popper et al. 2005, 2007). While experiencing TTS, fishes may have a decrease in fitness in terms of communication, detecting predators or prey, and/or assessing their environment.

Masking is a hearing impairment with respect to the relevant sound sources normally detected within the soundscape (e.g., See Chap. 3.4 on Masking). However, the consequences of masking for fishes and sea turtles have not been fully examined. It is likely that increments in background sound within the hearing bandwidth of fishes and sea turtles may render the weakest sounds undetectable, render some sounds less detectable, and reduce the distance at which sound sources can be detected. Energetic and informational masking may increase as sound levels increase, so that the higher the sound level of the masker, the greater the masking. Setting guidelines for masking is difficult because of spatial and other releases from masking that are not yet fully known for fish, but are well known for mammals and birds.

Table 7.1 presents definitions for these Guidelines. Effects on hearing include TTS and masking. TTS is defined herein as any reduction in hearing threshold of 6 dB or greater in terms of sound pressure or particle motion. This level is selected since changes of less than 6 dB are generally difficult to differentiate experimentally.

In these Guidelines, masking is considered to be the impairment of the ability to detect sounds, including the auditory scene, by a reduction in signal to noise ratio of greater than 6 dB. This level is suggested since it is unlikely that a change of less than 6 dB is measurable experimentally in fishes or turtles. Also, masking only occurs as long as the masker is present. A very short period of masking, such as from a single pulse of sound, or widely separated pulses, may not affect fitness. However, if impulsive sounds are generated repeatedly by many sources over a wide geographic area then there is the possibility that the separate sounds may merge and the overall background noise be raised (e.g. Nieukirk et al. 2004).

No studies have been conducted on hearing loss or effects of exposure to intense sounds on hearing in any turtles. Moreover, there have been no studies to determine if the hair cells in the basilar papilla of turtles are lost during exposure to intense sounds or if turtles can recover hair cells lost through exposure to intense sounds.

5.3 Effects on Behavior

The National Research Council (NRC 2005) discussed the possible effects of sound upon behavior, including communication between conspecifics and detection of predators and prey. In its report, the NRC states that an action or activity becomes biologically significant to an individual animal when it interferes with normal behavior and activity, or affects the animal's ability to grow, survive, and reproduce. Such effects may have consequences at the population-level and may affect the viability of the species (NRC 2005).

A significant issue in understanding the effects of man-made sound on fishes is that studying their responses to sounds is difficult and costly to perform, especially in the field. Many factors may influence the results, and a careful approach based on well-designed experiments must be adopted. Behavioral studies on captive or caged animals have been conducted (e.g., McCauley et al. 2000; Popper et al. 2007; Sarà et al. 2007; Mueller et al. 2010; Fewtrell and McCauley 2012) but results with free-living animals may differ because of the many subtle factors that determine an animal's behavior in a natural setting.

A range of responses has been observed when the behavior of wild fishes has been studied in the presence of man-made sounds. Some fishes have shown changes in swimming behavior and orientation, including startle reactions (Pearson et al. 1992; Wardle et al. 2001; Hassel et al. 2004). The response may habituate with repeated presentations of the same sound. Sound can also cause changes in schooling patterns and distribution (Pearson et al. 1992). For example, the horizontal and vertical distributions of both pelagic and demersal fishes were altered during and after airgun operations (e.g., Dalen and Knutsen 1987; Engås et al. 1996; Engås and Løkkeborg 2002; Slotte et al. 2004; Løkkeborg et al. 2012a, b). In some circumstances, fish react to approaching ships, leading to concern by fisheries scientists that vessel avoidance will bias stock-assessment surveys by research vessels (Mitson and Knudsen 2003). Fish can respond to approaching vessels by diving towards the seafloor or by moving horizontally out of the vessel's path, with reactions often initiated well before the vessel reaches the fish (Vabø et al. 2004; Ona et al. 2007). However, the stimuli to which these fishes are responding are not always clear (Sand et al. 2008).

Researchers have also attempted to monitor sea turtle avoidance to sound during an active seismic survey (Weir 2007; DeRuiter and Doukara 2010). Weir (2007) observed fewer sea turtles near airguns as they were firing (as opposed to silent airguns). However the source of agitation could not be identified; the turtles may have reacted to the ship and towed equipment rather than specifically to the airgun (Weir 2007).

Reductions in fish catches have been observed in commercial line and trawl fisheries during and after seismic surveys (Skalski et al. 1992; Løkkeborg and Soldal 1993; Engås et al. 1996; Løkkeborg et al. 2012a, b). In some studies catch rates in static gillnets increased, and this result was attributed to a rise in swimming activity by the fish in response to airgun sounds, thus making the fish more vulnerable to capture by gillnets (Løkkeborg et al. 2012a, b).

For the purposes of these Guidelines (Table 7.1), effects on behavior refer to substantial changes in behavior for a large proportion of the animals exposed to a sound. This may include long-term changes in behavior and distribution, including moving from preferred sites for feeding and reproduction, or alteration of migration patterns. This behavioral criterion does not include effects on single animals or small changes in behavior such as a startle response or minor movements.

5.4 Population-Level Effects on Fitness and Survival

From a conservation perspective, the immediate impact of man-made sounds on individuals or on schools of fish is less important than the long term impact on populations and ecosystems, either alone or in combination with other stresses (which will often include fishing). A reduction in the numbers of fish through exposure to sound may or may not have a measurable effect on fish population recruitment. Some fish populations go through a period of density-dependent mortality, and removing a small number of animals may simply result in their replacement through the improved survival of others. Nevertheless, a reduction in spatial distribution of fish or a reduction in genetic diversity could, by disrupting the migrations of a particular sub-population, have serious consequences at a local level. There have been attempts to develop predictive models for marine mammals based on studies of the disruption to individuals and examining the effects on key life functions like feeding, growth, reproduction and migration (NRC 2005). But such models are still in their infancy and are unlikely to be broadly applicable even for marine mammals, much less for fishes and sea turtles.

While individual organisms may be of major consequence from a conservation and regulatory perspective when considering marine mammals (Southall et al. 2007) and sea turtles, the greater concern with regard to fishes is with the impact upon populations. While death of a single fish may have some importance if it is a protected species, it may be more relevant to consider effects upon populations or sub-populations. As noted above, such population effects could result if exposure to sound has an impact on breeding or feeding success or alters migratory patterns. Widespread impaired hearing or sub-lethal injuries may result in impacts on populations. Effects could last from the immediate period of sound exposure (e.g., masking) to several weeks if fish are displaced from their preferred areas (Engås et al. 1996; Slotte et al. 2004; Løkkeborg et al. 2012a, b).

Chapter 6
The Nature of Man-Made Sound

The sounds originating from different sources may differ. The sound from a single source may change during propagation, and the signal received by the animal may differ from the sound close to the source (Ellison and Frankel 2012). Propagating through water and through the substrate may change the characteristics of a sound. Short, abrupt sounds may become lengthened and their onset smoothed as a result of their transmission over long distances due to the effects of absorption, multipath transmission, modal dispersion, and refraction, while repeated sounds and their echoes can merge together to become more continuous (Nieukirk et al. 2004). While there are a variety of models that help understand propagation, it is important to keep in mind that sound propagation models designed for deep ocean environments will not be appropriate in shallow water environments where the wavelength of the sounds may approach the depth of the water and a modal or full wave model is required (Hovem et al. 2012).

A number of other factors must also be taken into consideration. For example, the sound field impinging on an animal may differ depending on the animal's position relative to the seabed (Hazelwood 2012; Hazelwood and Macey, 2015). Animals near the seabed may not only detect water-borne sounds, but also sound that propagates through the substrate and re-enters the water (e.g., as in pile driving and seismic exploration). Moreover, animals may be affected by the particle motion component of the sound field as well as the sound pressure, and near-field effects may increase their exposure to particle motion.

In these Guidelines, a distinction is made between transient and continuous sounds. Transient sounds are short-lived and can be impulsive or non-impulsive. Impulsive sounds are typically abrupt, of brief duration, and may contain a wide range of frequencies. Examples are the sounds made by explosions, seismic airguns, and pile driving. Near their source, such sounds often have a rapid rise time, quickly reaching a maximum value, followed by a period of decay. With increasing distance from the sound source the time structure becomes drawn out and less impulsive in character.

Sounds from explosions can be especially abrupt, with very short rise-times. Sounds from seismic airguns and pile driving are similar to one another, but they have

A.N. Popper et al., *ASA S3/SC1.4 TR-2014 Sound Exposure Guidelines for Fishes and Sea Turtles*, SpringerBriefs in Oceanography, DOI 10.1007/978-3-319-06659-2_6,

longer rise times than explosive sounds. Some transient sounds may lack the extreme properties of impulsive sounds, with much slower rise-times and longer durations. Examples include sounds from marine construction work and some types of sonar.

Some sounds, especially impulsive sounds, may be either single, like the sound from a single explosive charge, or repeated, like the sounds from seismic airguns or pile driving. With repeated sounds, the duty cycle, duration, and number of sounds need to be taken into account in assessing any effects upon animals.

Continuous sounds can be tonal (consisting of one or more frequencies, with or without harmonics), or broadband (containing a wide range of frequencies), and can change in amplitude with time. Some continuous sounds may be 'rougher' than others, with a relatively high crest factor or kurtosis (Henderson and Hamernik 2012). Examples of sources producing continuous sounds include ships, aircraft, dredgers, machinery operations such as drilling or wind turbines, and some active sonar systems.

With all sounds it is necessary to consider not only the frequency composition but also the temporal structure. Sounds from some sources may rise in amplitude as the source approaches and then fall as the source moves away from the receiving animal. Examples include the continuous sounds from ships, or the impulsive sounds from seismic airguns. Other sources like pile drivers may be stationary, but the receiving animals themselves may be moving, resulting in the received amplitude changing with time.

With such wide variations in sound sources and changes in the characteristics of sounds as they propagate away from the source, it is necessary to employ a range of metrics to describe sounds fully. These metrics are discussed below and also by Southall et al. (2007), Ellison and Frankel (2012), and Ellison et al. (2012).

6.1 Metrics

6.1.1 Sound Pressure

In general, the metrics to be used in characterizing continuous sound in terms of its risk of damage to fishes and sea turtles are the root-mean-square (rms) sound pressure, the peak sound pressure, and, in some cases, the corresponding particle motion (Chap. 6.1.2) in three dimensions. Sound levels are usually given as the decibel (dB) value with respect to a reference value. For underwater sound the sound pressure reference value used is 1 micropascal (μPa).

6.1.2 Particle Motion

Particle motion (displacement, velocity, and acceleration) stimulates the otolithic organs within the ears of fishes. The measurement and analysis of this feature of all acoustic fields must be undertaken in any study of fish hearing since it is likely to be the major source of sound stimulation to a majority of fish species (Chap. 3.1).

Acoustic particle motion is defined as that motion caused by a sound wave of a given infinitesimal part of the medium relative to the medium as a whole, and it is an integral part of any sound field. Unlike pressure, particle motion is directional in nature and is typically described using three dimensional vector notation. In the far field of a sound source (often defined as starting at a distance of about one-sixth the wavelength of a sound from the source – van Bergeijk 1964), and away from any boundaries, the particle velocity is directly related to the pressure as a ratio proportional to the acoustic impedance of the medium, and the predominant vector is in the direct path from the source. Within the near field of a sound source, the ratio of particle velocity to pressure increases with diminishing distance to the source. The general relationship between pressure and velocity in both the near and far field of a source was described by Siler (1969) who summarized earlier work (e.g., Harris and van Bergeijk 1962; Harris 1964; van Bergeijk 1964). The three dimensional particle motion field is quite complex near boundaries that include the air/water interface and the seabed, as well as in shallow water (Parvulescu 1964; Rogers et al. 2015). In these instances the particle motion is often unpredictable, and should be measured. Particle motion levels can be given as rms or peak values expressed in dB with respect to a reference value of displacement, velocity, or acceleration. While various reference values have been used in the literature, efforts toward standardization are being made by the International Organization for Standards in ISO/DIS 1683 (2013). This standard recommends the following: 1 pm (picometer) for sound particle displacement, 1 nm/s for sound particle velocity, and 1 μm/s^2 for sound particle acceleration.

6.1.3 Peak Levels

Peak sound pressure (or particle motion) is the maximum absolute value (either positive or negative) of the instantaneous sound pressure (or motion) during a specified time interval and is properly denoted as p_{max}. Peak is a useful metric for characterizing impulsive sounds. Peak-to-peak is the difference between the absolute value of the maximum negative and positive instantaneous peaks of the waveform. Positive and negative peak pressures may have different effects – negative pressures result in expansion and cavitation while positive pressures result in compression. In measuring peak levels, sampling must be at or above the Nyquist frequency (Nyquist 2002) to prevent the actual peak from falling between adjacent sampling points.

6.1.4 Sound Intensity

Sound *amplitude*, as characterized by various sound-pressure and particle-motion metrics described in Chap. 6.1.2 should not be confused with sound intensity, which is defined as the time-averaged power per unit area (Coppens et al. 2000) in watts per meter square. The sound intensity corresponds to local transport of sound energy and is related to the product of the sound pressure and the particle velocity component in-phase with the sound pressure. Most laboratory studies involve complex sound fields that have spatially varying and frequency-dependent relationships between pressure and velocity. In these circumstances, sound intensity cannot be estimated from individual pressure measurements, but requires measurements of the pressure gradients using hydrophone arrays or a direct measurement of the particle motion using orthogonally orientated, neutrally buoyant, accelerometers.

6.1.5 Impulse

It has been recognized that neither rms nor peak levels are sufficient for characterizing the total energy or temporal characteristics of single impulsive sounds such as those generated by explosions, pile driving strikes or the discharge of seismic airguns. Impulse in this context is the time integral of pressure through the waveform expressed in Pa·s. Impulse is sometimes alternatively defined as the sum of the pressure over the duration of the waveform.

6.1.6 Sound Exposure Level

Another relevant metric is the sound exposure level (SEL), the time integral of the pressure squared for an event, which is an index of the total energy in a sound. SEL is usually expressed in dB re 1 $\mu Pa^2 \cdot s$. In the far field, the SEL is proportional to the total acoustic energy of the pulse.

However, the relationship between pressure and particle velocity is generally assumed to be that for a plane wave (in which the acoustic energy flux density or intensity is directly proportional to p^2). In many underwater environments, where the relationship between acoustic pressure and particle velocity is more complex, the total energy flux density will *not* be equivalent.

For pulses and transient sounds, the acoustic energy flux density enables sound exposures of differing duration to be related to one another for purposes of assessing exposure risk. However, it should be noted that the SEL metric becomes less useful, and possibly misleading, at longer exposure durations (Finneran and Schlundt 2010).

The SEL metric also enables the integration of sound energy across multiple exposures from sources such as seismic airguns, pile driving, and most sonar signals. Accordingly, SEL serves as an index for accumulated sound energy.

6.1.7 Cumulative Energy (Cumulative Sound Exposure Level)

A significant issue of concern is the accumulation of sound energy from repeated impulsive sounds. If an animal is exposed to repeated sounds, such as multiple pile strikes or repeated pulses from a seismic airgun, effects may be a function of the energy in all the sound events accumulated over time.

Recent studies of pile driving have often referred to SEL in terms of the single strike (SEL_{ss}) as an index of the energy when the pile is struck once. In addition, the SEL has also been used as an index of the total energy over the duration of the sound-making activity (e.g., the period of driving of an individual pile or the passing of a seismic array). In this case, the SEL is designated as the cumulative SEL (SEL_{cum}) (Hastings and Popper 2005; Popper and Hastings 2009; Halvorsen et al. 2011, 2012a; Casper et al. 2012b).

SEL_{cum} is the linear summation of the individual sound events over the time period of interest. If the SEL_{ss} is approximately the same for all events, then the SEL_{cum} can be estimated as $SEL_{ss}+10\log_{10}(N)$, where N is the number of impulsive events. The period over which the SEL_{cum} is accumulated must be carefully specified, however. For example, SEL_{cum} may be defined over a standard period (e.g., 12 hours of pile driving) or for the duration of an activity (e.g., the full period of construction), or over the total period that the animal will be exposed. Whether an animal would be exposed to a full period of sound activity will depend on its behavior, including whether it stays in the vicinity of the sound or moves away. Movement of the source itself will also have an effect.

Thus, complete characterization of SEL_{cum} should include not only the number of sound events, but also the time period over which the summation is performed, the distribution of sound events within that period, and changes in the magnitude of the individual sound events. Both the magnitudes of the individual sound events and the number of events will influence effects upon animals. The effects on the tissues of a fish may be very different depending on the time between strikes, with more tissue recovery possible with longer periods between sounds. At this time, nothing is known about how different inter-strike intervals might affect animals, and the lack of such data makes it hard to provide full guidance on effects of impulsive sounds.

There are some data on the effects of impulsive sounds on fishes (Chap. 7.1), and these data are instructive in developing initial guidelines. Importantly, while the studies were done relative to pile driving, the similarity of pile driving to other impulsive signals (e.g., from airguns) suggests that the results reported are relevant for other impulsive sources.

6.1.8 *Frequency Weighting*

The hearing of different species is frequency-dependent. Rather than express received sound pressures in terms of their levels over a broad bandwidth, levels can be weighted by the frequency response of hearing for the animal. This approach has been widely used for examining exposure of humans to sound (e.g., Henderson and Hamernik 2012).

Southall et al. (2007) developed M frequency weighting functions for five functional hearing groups of marine mammals, with the same mathematical structure as the C frequency weighting used in human hearing. This method was designed specifically to represent the broader bandwidth at which auditory effects (TTS and/or Permanent Threshold Shift - PTS) might occur for higher exposures.

A similar attempt at weighting has been made for individual species of fish and other animals by Nedwell et al. (2007) (see also Chap. 7.2). Their dBht (Species) metric expresses the level of received sound pressure weighted by a filter that reflects the frequency-dependent sensitivity of hearing for the species of interest.[1] The dBht is not an absolute sound level unit. Rather, it is said by Nedwell and his colleagues to provide a prediction of the perceived loudness of the sound to the animal.

The application of weighting requires reliable measures of hearing sensitivity versus frequency (audiograms), but these are only available for a few fish species. As discussed in Chap. 3.1, confidence in the validity of audiograms for many species is limited because of the poor acoustic conditions surrounding the experiments, uncertainties as to whether particle motion or pressure is the relevant sound dimension, and the methodologies applied to determine thresholds.

One of the (many) problems with the dBht concept is that while a large proportion of fish species are primarily detecting particle motion, the audiograms used by Nedwell et al. (2007) to determine dBht (Species) levels are expressed in terms of sound pressure and are not appropriate for these species. A second concern is that the hearing sensitivities for some of the species discussed by Nedwell et al. (2007) were derived using AEP measures and not behavioral measures. Behavioral and AEP thresholds may not correlate well, particularly in estimating threshold sensitivity (Ladich and Fay 2013; Sisneros et al. 2015). Rather than apply a weighting function species by species, it may be more appropriate to apply generalized weighting functions for the functional hearing categories, defined in Chap. 4.4.

In summary, while the general concept proposed by Nedwell et al. (2007) may have some value in the context of behavioral responses by fish, its application and adoption requires far more scientific validation and the inclusion of those species that primarily respond to particle motion. Furthermore, caution is needed in applying weighting to sounds that are potentially injurious. Sounds outside the hearing range of the animals, that are inaudible, may be capable of causing damage to tissues. In particular, the high frequencies associated with rapid rise-times may bring about or exacerbate injury. For these reasons it would be premature to apply any weighting in the development of guidelines for fish and turtles.

6.2 Sources

6.2.1 Explosions

In-water explosions produce a spherical shock wave that travels at speeds greater than the speed of sound in water. A large oscillating gas bubble is also produced that radiates sound. Near the source, the pressure rise-time for some explosives,

---

[1] We note that it is inappropriate to add a subscript to a dimensional unit like the dB, and that such use has been disallowed by ISO 80000-8:2007 (2007). However, it has been done by Nedwell et al. (2007) and the notation must be included here in order to enable discussion of the idea.

such as TNT, is nearly instantaneous, followed by exponential decay. The rise time for other explosives may be longer, and the decay of the pulse slower (Urick 1983). This rise time affects the frequency content in the signature of the explosion, with longer rise times lacking the higher frequencies.

Underwater explosions have been extensively studied and are described by Cole (1948) and Urick (1983). However, although the physical principles of underwater detonations are well understood, there are significant gaps in applying this knowledge in specific situations, like the use of explosives for the removal of offshore structures (Continental Shelf Associates 2004). These knowledge gaps inhibit the use of various biological effects (fish kill) models, such as the "impulse metric" model of Yelverton et al. (1975) and the "bladder oscillation parameter" model of Goertner (1978) to predict the risk to fish and other animals from the pressure wave generated by an explosion.

Explosions are typically described by metrics that specify the amplitude, energy, and time-space related characteristics of the pressure wave. Those metrics may include the peak pressures, both positive and negative, the impulse, and the SEL. The levels of particle motion are potentially relevant for explosions as they may be very high. The rise-time may also be relevant since it can be very rapid and may be important in causing tissue damage.

6.2.2 Seismic Airguns

Airguns are used as an acoustic source for seismic exploration to survey the substrate. Airguns work by producing an air bubble from a compressed air supply. The air bubble first expands rapidly and creates an impulsive signal, but with a slower rise time to peak pressure than occurs with explosions. The bubble creates a series of smaller oscillations that follow the primary pulse created by the initial formation of the bubble. The sound impulse generated by a single airgun is omnidirectional, with peak energy at low frequencies, typically 20-50 Hz with declining energy at frequencies above 200 Hz. Arrays consisting of many airguns are towed behind vessels during a seismic survey. The array is designed to maximize the initial pulse and minimize the effect of bubble oscillations and surface reflections. During the seismic survey, the array is fired at regular intervals (e.g., every 10 to 16 seconds) as the towing vessel moves ahead, and the survey may continue in the same general region for hours or days (or longer). The sound from the array is directed downwards to enter the seabed and the reflected sound is detected by hydrophones towed behind the vessel (Caldwell and Dragoset 2000; Mattsson et al. 2012).

The impulsive sound generated beneath the airgun is the result of both the direct pulse and a very strong reflected pulse from the sea surface. The source level of an airgun array can vary considerably with the design of an array and the type of airguns (Richardson et al. 1995; MacGillivray and Chapman 2005; OGP 2011). Source levels are often estimated from measurements made at a considerable distance from the array and extrapolated back to 1 meter from a hypothetical point source. There are difficulties in applying this method to a large distributed source such as an airgun array. Sound levels at an animal directly below, or close to, the seismic array can be overestimated by 20 dB or more by such an extrapolation.

A study of ambient sound in the North Atlantic (Nieukirk et al. 2004) showed that airgun activity contributes significantly to ocean sound levels and can appear to be more continuous than impulsive because of reverberation from the surface and seabed. Dragoset (2000), Laws and Hedgeland (2008), Tashmukhambetov et al. (2008), and Laws (2012) discuss the acoustic characteristics of airgun arrays.

6.2.3 Pile Driving

Pile driving is commonly used for the construction of foundations for a large number of structures including bridges, buildings, retaining walls, harbor facilities, offshore wind turbines, and offshore structures for the oil and gas industry (e.g., Reyff 2012). It always involves multiple strikes over an extended period of time, with an average strike interval of 1.0 to 1.5 seconds.

Pile driving can be a source of underwater sound if the pile being driven is in water or on land near water. The substrate can contribute either via direct propagation or via interface (Sholte) waves (Hazelwood 2012; Hazelwood and Macey, 2015). The latter originate at the water-sediment interface and have large particle velocity components that decay rapidly with distance from the interface (Brekhovskikh and Lysanov 1982). Shear waves and interface waves travel slower than sound waves within the substrate and their peak energy is at lower frequencies (Dowding 2008).

The impulsive sounds generated by impact pile driving are characterized by a relatively rapid rise time to a maximal pressure value followed by a decay period that may include a period of diminishing, oscillating maximal and minimal pressures (Illingsworth and Rodkin 2001, 2007; Reyff, 2012). Peak levels resulting from impact pile driving vary substantially and depend on numerous factors such as pile type and diameter, hammer size, substrate, etc. The predominant energy in pile impact impulses is at frequencies below 500 Hz (Laughlin 2006; Reyff 2008, 2012).

Peak sound pressure levels are useful for characterizing pile driving strikes but do not account for the total energy of the sound. The sound exposure level (SEL), as described above, is related to the total acoustic energy of the pulse and enables sound exposures of differing duration to be related to one another for purposes of assessing exposure risk.

Vibratory pile driving produces a continuous sound with peak pressures lower than those observed in impulses generated by impact pile driving. Sound signals generated by vibratory pile driving usually consist of a low fundamental frequency characteristic of the speed of rotation of the revolving mass in the vibratory hammer (typically on the order of 30 Hz), and its higher harmonics (e.g., Laughlin 2006).

6.2.4 *Active Sonar*

Active sonar and echo sounders are in operation throughout the world's oceans as well as in freshwater lakes and rivers. The primary sonar characteristics that vary with application are the frequency band, signal type (pulsed or continuous), rate of repetition, and source level. They can be roughly divided into three categories depending on their primary frequency of operation; low frequency (LF) for 1 kHz and less, mid frequency (MF) for 1 kHz to 10 kHz, and high frequency (HF) for 10 kHz and greater. Low, and possibly mid, frequency sonars are most relevant to fishes and sea turtles because of the low frequency hearing ranges of these animals (e.g., Popper et al. 2007; Halvorsen et al. 2012b). Sonar usually operates with duty cycles (transmission time/total time) below 10 to 20% and with generally brief durations. However, multipath propagation can often be substantial for many of these systems, effectively prolonging the sonar sounds well beyond their nominal durations.

Low frequency systems are designed for long-range detection. For example, the U.S. Navy SURTASS LFA (low frequency active) system is described by Friedman (2006) as a vertical line array (VLA) of 18 elements operating between 100 and 500 Hz. Signals projected include combinations of swept frequency (FM) and tones pulses, totaling up to 100 s in length with individual signals of the order of 10 s. The interval between transmissions varies between 6 and 15 minutes.

6.2.5 *Continuous Sound Sources*

The most common and best-studied continuous sounds in the oceans are those produced by ships as well as smaller vessels, including pleasure craft. However, continuous sounds are also produced by other sources, such as vibratory pile drivers and vessels dredging for aggregates (Robinson et al. 2011). In addition, over long distances, emissions from seismic surveys or impact pile driving may appear to be continuous as a result of multipath effects.

The sounds of vessels are predominately low frequency (i.e., below 1 kHz) from onboard machinery, hydrodynamic flow around the hull, and from propeller cavitation, which is typically the dominant source of noise (Ross 1987, 1993). Radiated vessel noise relates to many factors, including ship size, speed, load, condition, age, and engine type (Richardson et al. 1995; Arveson and Vendittis 2000; NRC 2003; De Robertis et al. 2012; McKenna et al. 2012). Estimated source levels can range from < 150 dB to over 190 dB (re 1 μPa at 1m) for the largest commercial vessels (Richardson et al. 1995; Arveson and Vendittis 2000; Hildebrand 2009; McKenna et al. 2012).

Low frequency sounds from larger vessels can travel hundreds of kilometers and can increase ambient noise levels over large areas of the ocean, interfering with sound communication in species using the same frequency range (see Southall 2005).

Tens of thousands of large commercial vessels are typically underway at any point in time, concentrated in high-traffic sea-lanes and ports, and presenting an effectively continuous noise source in many parts of the ocean.

The number of large commercial ships has doubled between 1965 and 2003 to nearly 100,000. Shipping industry analysts have forecast that the volume of cargo shipped will again double or triple by 2025, with an expected attendant increase in the amount of ambient noise entering the ocean from commercial shipping (NRC 2003). In much of the northern hemisphere, shipping noise is the dominant source of underwater noise below 300 Hz (Ross 1987, 1993). Vessel operations have increased over time and as a result have increased low-frequency ambient noise levels in some areas (see Curtis et al. 1999; Andrew et al. 2002; NRC 2003; McDonald et al. 2006). One of the most serious implications of this increase in shipping noise is the impact it may have in masking sounds of biological importance, including sounds made by fishes. ANSI/ASA S12.64-2009/Part 1 standard (2009) and ISO/PAS 17208-1:2012 (2012) have been formulated and have established templates for the measurement of vessel noise.

Chapter 7
Sound Exposure Guidelines

7.1 Current Guidelines for Injury

Currently there are no U.S. domestic or international standards for exposure of fish to impulsive sounds. However, the National Marine Fisheries Service (NMFS) has developed interim criteria for pile driving (for use primarily on the west coast of the United States) that use dual criteria (FHWG 2008; Woodbury and Stadler 2008; Stadler and Woodbury 2009; Caltrans, 2009). These criteria specify both a maximum permitted sound pressure level for a single pile driving strike and a maximum accumulated sound exposure level for lower level signals. The SPL_{peak} was selected to be 206 dB re 1 μPa and the maximum SEL_{cum} was designated as 187 dB re 1 $\mu Pa^2 \cdot s$ for fish ≥2 grams and 183 dB re 1 $\mu Pa^2 \cdot s$ for fish <2 grams. The idea of dual criteria (Carlson et al. 2007) was formulated to ensure that fish were neither exposed to single strikes at very high peak levels nor to high levels of accumulated energy from repeated impulsive sounds. The levels for both peak and cumulative energy selected by NMFS were largely based on data from the mortality of fishes exposed to explosives (Popper and Hastings 2009).

Recently, Halvorsen et al. (2011, 2012a, c) described an extensive investigation into the effects of simulated pile driving impulsive sounds upon juvenile Chinook salmon as well as on Nile tilapia (*Oreochromis niloticus*), hybrid striped bass (*Morone chrysops* x *Morone saxatilis*), lake sturgeon (*Acipenser fulvecens*), and hogchoker (*Trinectes maculates*) (Casper et al. 2012b, 2013a, b; Halvorsen et al. 2012c). An important advance with this work was the development of an injury response variable, the Response Severity Index (RSI), based on the physiological impact of different injuries on the health of the fish (see Chap. 5.1). Halvorsen and her collaborators were then able to determine peak sound pressure levels associated with different levels of the RSI for different species. They also derived a function that related both the SEL_{cum} and the number of pile strikes to the RSI (Fig. 7.1).

An important finding from these studies was that tissue damage increased both as energy accumulated over multiple strikes (SEL_{cum}) and/or as the energy in single strikes increased. However, the relationship was not linear, and it is clear that an

A.N. Popper et al., *ASA S3/SC1.4 TR-2014 Sound Exposure Guidelines for Fishes and Sea Turtles*, SpringerBriefs in Oceanography, DOI 10.1007/978-3-319-06659-2_7,

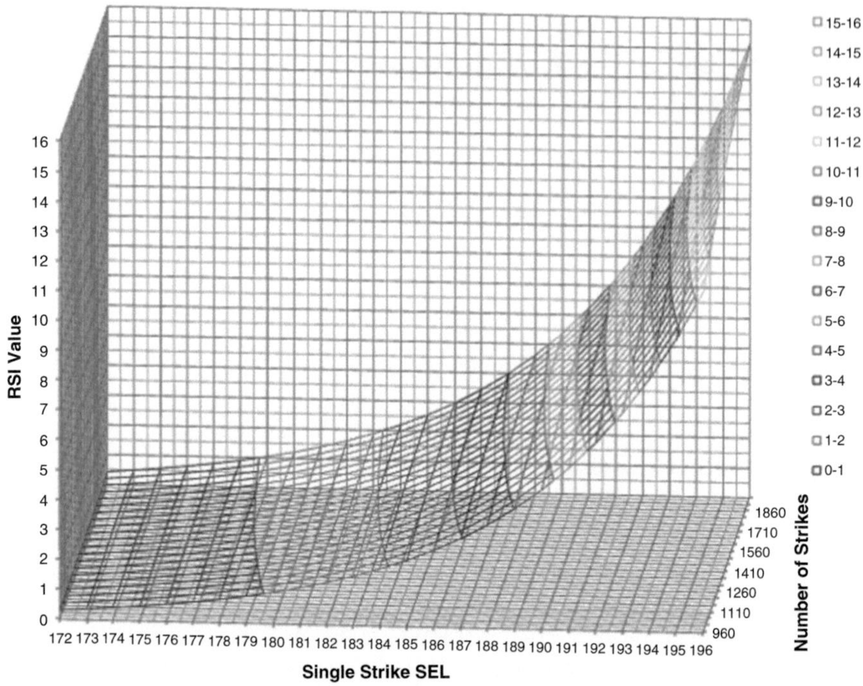

Fig. 7.1 Response Severity Index (RSI) as a function of SEL_{ss} and number of strikes. RSI increases with energy in the single strikes of a pile by a pile driving hammer and with the number of strikes taken to drive a pile

"equal energy hypothesis," in which effects are related to total energy no matter how that total energy was achieved, is not correct (Halvorsen et al. 2011, 2012a).

The development of impulsive sound exposure criteria for pile driving is based on the data provided by Halvorsen et al. (2011, 2012a) and requires that a level of injury severity be selected as a starting point. The next step is specification of the levels for individual impulsive sounds that do not exceed that level of injury for a single strike. That specification could be in terms of SPL_{peak} or SEL_{ss}, as these are highly correlated for pile driving sounds (Carlson and Weiland 2007). Since there is also concern for effects of multiple strikes where no single strike approaches the SPL_{peak}, the final step in the development of criteria is to define an SEL_{cum} which is based on the combination of SEL_{ss} and number of strikes that would result in the onset of the lowest level of injury (RSI) that would be considered deleterious to the species of concern.

Can these criteria, derived from experiments simulating pile driving, be applied to other impulsive sources, such as seismic airgun arrays? While the acoustics of the impulsive sounds generated by driving piles and firing seismic airguns are similar, the basic nature of exposure for a fish differs between the two sources. Current pile driving exposure criteria assume that both the source and the exposed fish are stationary over the duration of driving the pile, although there may be scope for movement by many species of fish. However, seismic arrays are towed at a speed of about 4 knots, meaning that an array would move about 33 m between shots given a shot frequency of once every 16 seconds. A considerable distance may be traveled over the full seismic survey. As a consequence, the direct application of cumulative criteria developed for pile driving to seismic airguns would not be appropriate since the received level of each shot at the fish would always be at a different SEL_{ss}.

7.2 Current Guidelines for Behavior

Guidelines on the impact of impulsive sounds upon the behavior of fishes and turtles are very limited. The U.S. NMFS currently uses a criterion for behavioral response of 150 dB re 1 μPa (Stadler and Woodbury 2009), but it is not clear whether this is a peak or rms level. However, as pointed out by Hastings (2008), no one is sure of the origin of this number, and it is not clear if it has any scientific validity. Moreover, the criterion does not specify a particular behavior, but simply assumes there is the potential to experience a behavioral response.

Nedwell et al. (2007) proposed a set of guidelines for behavior utilizing the dBht (Species) concept (Chap. 6.1.8). They suggested that the following dBht levels above threshold elicited particular responses: 0–50 dB elicits a mild reaction in a minority of individuals, probably not sustained; 50–90 dB elicits a stronger reaction by the majority of individuals, but habituation may limit effect; 90 dB and above elicits a strong avoidance reaction by virtually all individuals; above 110 dB is the tolerance limit of sound; unbearably loud.

These figures were largely derived from controlled exposure experiments with a fish avoidance system at a power-plant water intake (Maes et al. 2004), supplemented by observations from the testing of a fish guidance system in shallow raceways (Nedwell et al. 2007). Few independent experiments have been carried out to confirm these values for other species or in other circumstances, especially at sufficiently high sound levels to determine how fish respond at 90 dB or more above their hearing threshold.

Beyond the limited experimental data supporting the guidelines proposed by Nedwell et al. (2007), there are a number of other issues that preclude the use of these guidelines (beyond the overall critique described in Chap. 6.1.8). Most importantly, it is very clear to anyone working with animal behavior that one number for a guideline or criteria can never fit all fishes, since species vary greatly in so many ways. There are differences in their hearing capabilities and how they respond to stimuli in general (swim away, bury in the substrate, etc.) that will affect whether a sound at a given level will elicit a response or not. Moreover, responses to a signal may vary within a species, and even a single animal, depending on things such as sex, age, size, and motivation (feeding, mating, moving around a home range, etc.)

As a consequence, developing behavioral guidelines is far harder than developing guidelines for physiological effects. Carrying out appropriate experiments that take into consideration factors that range from species differences to motivational state of an animal is very complex. This difficulty explains why there are very few studies of behavior of wild fish or turtles. Animals in tanks or even in large enclosures show very different responses to behavioral stimuli than do wild animals (e.g., Oldfield 2011). Studies on captive animals are suitable for gaining physiological information such as hearing sensitivity, but not for understanding how a wild animal will respond behaviorally to a stimulus.

7.3 Practical Approach to Recommending Sound Exposure Guidelines

For the purposes of this report, guideline levels represent the received level of exposure to sound that produces a specified effect based on current data. Sounds above the guideline level are considered likely to result in that effect, with higher sound levels likely to produce greater effects. The actual guideline levels will not necessarily be the same for all sound sources or all species, or even for the same source and species under different circumstances. Guidelines for different types of sound sources are only provided where data exist on received sound levels.

Table 7.1 Definition of Effects Used in Guidelines Tables

- *Mortality and mortal injury* – immediate or delayed death.
- *Recoverable injury* – injuries, including hair cell damage, minor internal or external hematoma, etc. None of these injuries are likely to result in mortality.
- *TTS* – short or long term changes in hearing sensitivity that may or may not reduce fitness. TTS, for these Guidelines, is defined as any change in hearing of 6 dB or greater that persists. This level is selected since levels less than 6 dB are generally difficult to differentiate. It is also the view of the WG that anything less than 6 dB will not be a significant effect from the standpoint of hearing.
- *Masking* – impairment of hearing sensitivity by greater than 6 dB, including all components of the auditory scene, in the presence of noise.
- *Behavioral effects* – substantial change in behavior for the animals exposed to a sound. This may include long-term changes in behavior and distribution, such as moving from preferred sites for feeding and reproduction, or alteration of migration patterns. This behavioral criterion does not include effects on single animals, or where animals become habituated to the stimulus, or small changes in behavior such as a startle response or small movements.

The relative risk of an effect taking place is indicated as being "high," "moderate," and "low."

To determine where sufficient data exist for establishing valid sound exposure guidelines, several classification schemes were developed for organizing data from the scientific literature. First, fishes and sea turtles were grouped into the five classes based on auditory structures described in Chap. 4. Second, man-made sound sources were grouped into five classes as discussed in Chap. 6.2.1 through 6.1.5. Finally, the effects of sound exposure were placed into five categories as listed in Table 7.1, recognizing that sound exposure might result in single effects or a combination of any number of effects.

These classification schemes are qualitative and descriptive and based on the best data available at the time of this writing. Moreover, since there are often no data to fill many of the table cells, data have been extrapolated between sources and effects where it was considered appropriate. Clearly, it will be necessary to refine all of the guidelines over time as additional data and more sophisticated means of assessment become available.

7.4 Explanation of the Analyses

Sound exposure guidelines are provided for *received sound levels*. Where data exist that can be used to suggest provisional guidelines, received signal levels are reported in appropriate forms (e.g., peak, rms, SEL). Accompanying each Table is a discussion of the source of the guidelines provided. Cells with available data are shaded to make them stand out. The best available data have been used for either the sources of interest, or for similar sources. Thus, the data in the tables for pile driving and seismic airguns use the same dataset since the sources yield sounds that are similar in their spectra, rise times, and duration.

Where insufficient data exist to make a recommendation for guidelines a subjective approach is adopted in which the relative risk of an effect is placed in order of rank at three distances from the source – near (N), intermediate (I), and far (F) (top to bottom within each cell of the table, respectively). While it would not be appropriate to ascribe particular distances to effects because of the many variables in making such decisions, "near" might be considered to be in the tens of meters from the source, "intermediate" in the hundreds of meters, and "far" in the thousands of meters.

The *relative* risk of an effect is then rated as being "high," "moderate," and "low" with respect to source distance and animal type. No assumptions are made about source or received levels because there are insufficient data to quantify what these distances might be. However, in general the nearer the animal is to the source the higher the likelihood of high energy and a resultant effect. In specifying these distances and the potential effects, regulators and others need to consider actual source and received levels and the sensitivity to the sources by the animals of concern. The rating for effects in these tables is highly subjective, and represents general consensus within the WG. However, these ratings are not hard and fast, and they are presented as the basis for discussion.

Finally, some cells are labeled "NA" (not applicable) if an effect of a particular type for a particular group of animals is considered not likely to happen.

7.5 Guidelines

7.5.1 Source: Explosions

For explosions (Table 7.2), the guidelines are based upon a single explosion from dynamite or another relatively small charge used to dismantle in-water structures. Different expectations would be indicated for larger or multiple explosions.

Table 7.2 Explosions: Guidelines for explosions. Levels other than for eggs and larvae from Hubbs and Rechnitzer (1952); levels for eggs and larvae from Wright and Hopky (1998). Guidelines are not provided for masking since the animals are not exposed to more than a few explosive events, and masking would not last beyond the period of exposure

Type of Animal	Mortality and potential mortal injury	Impairment			Behavior
		Recoverable injury	TTS	Masking	
Fish: no swim bladder (particle motion detection)	229 - 234 dB peak	(N) High (I) Low (F) Low	(N) High (I) Moderate (L) Low	NA	(N) High (I) Moderate (F) Low
Fish where swim bladder is not involved in hearing (particle motion detection)	229 - 234 dB peak	(N) High (I) High (F) Low	(N) High (I) Moderate (F) Low	NA	(N) High (I) High (F) Low
Fish where swim bladder is involved in hearing (primarily pressure detection)	229 - 234 dB peak	(N) High (I) High (F) Low	(N) High (I) High (F) Low	NA	(N) High (I) High (F) Low
Sea turtles	229 - 234 dB peak	(N) High (I) High (F) Low	(N) High (I) High (F) Low	NA	(N) High (I) High (F) Low
Eggs and larvae	>13 mm s^{-1} peak velocity	(N) High (I) Low (F) Low	(N) High (I) Low (F) Low	NA	(N) High (I) Low (F) Low

Notes: peak and rms sound pressure levels dB re 1 μPa; SEL dB re 1 $\mu Pa^2 \cdot s$. All criteria are presented as sound pressure even for fish without swim bladders since no data for particle motion exist. Relative risk (high, moderate, low) is given for animals at three distances from the source defined in relative terms as near (N), intermediate (I), and far (F).

Fishes: The effects of explosions on fish have been measured several times (e.g., Yelverton et al. 1975; Goertner et al. 1994; Keevin and Hempen 1997; Carlson et al. 2011). Govoni et al. (2003, 2008) examined the effects on eggs and larvae. For a discussion of the effects of underwater explosions upon fishes and other animals, and the models used to predict them, see Continental Shelf Associates (2004).

The water volume affected by the pressure wave generated by an explosion is complex and is a function of the depth of the explosion, the water depth, and the nature of the substrate. In general, two sub-regions within the total affected volume pose high risk to fish. One is in the immediate vicinity of the explosion where the compressive forces of the shock wave generated by the explosion predominate; the second is a more distant region where negative pressure resulting from the overshoot of the gas bubble created by an explosion and the reflection of the shock wave from the water surface can cause cavitation and negative pressures low enough to cause harmful expansion of swim bladders and other barotraumas, including mortality (Goertner 1978; Cole 1948). The rupture of the swim bladder, or the damage inflicted on other body organs, such as hemorrhage caused by rupture of veins or the occurrence of gaseous blood in the heart, may result in immediate or delayed death (McKinstry et al. 2007; Halvorsen et al. 2011, 2012a; Brown et al. 2012).

There is also evidence (e.g., Goertner et al. 1994; Stephenson et al. 2010; Halvorsen et al. 2012c) that little or no damage occurs to fishes without a swim bladder except at very short ranges from an in-water explosive event. Goertner (1978) showed that the range from an explosive event over which damage may occur to a non-swim bladder fish is on the order of 100 times less than that for swim bladder fish. However, it is not possible to clearly separate concussive and decompressive injuries to swim bladder fish because both concussion and decompression forces act near an explosion, although decompressive forces predominate with distance.

Yelverton et al. (1975) determined experimentally that the impulse magnitude required to produce 50% mortality increased with the mass of the fish. They also found that the degree of injury was the same in physostomous and physoclistous[1] fishes, presumably because the shock wave and any associated negative pressure pulse, particularly the inverted reflection of the high positive pressure shock wave from the water surface, has a very rapid rise time that prevents physostomous fish from adjusting their swim bladder volume before the onset of barotrauma.

Goertner (1978) concluded that the fish-injury predictive models based on impulse as the independent variable were only applicable when the explosion and fish were located at depths of less than 3.3 m. His data and model showed a decrease in susceptibility of fish to injury from explosions at greater depth. This decrease in susceptibility is a function of the decreasing difference between the hydrostatic pressure and the pressure generated by the explosion, as depth increases.

More recent studies of rapid decompression and exposure of fish to simulated pile driving impulsive sounds have clarified the importance of acclimation and acclimation depth in fish responses to decompressive forces (Stephenson et al. 2010; Halvorsen et al. 2011, 2012a). Acclimation is the process of fish achieving neutral buoyancy at a new depth by adjusting the volume of gas in their swim bladders. Non-acclimated swim bladder fish exposed to decompressive forces may respond like non-swim bladder fish, depending on the degree to which the swim bladder is inflated (Stephenson et al. 2010; Halvorsen et al. 2011, 2012a). These studies have also shown the importance of acclimation depth when considering the response of fish to negative pressures (decompression).

Yelverton et al. (1975) and others (Stephenson et al. 2010; Halvorsen et al. 2011, 2012a) provide evidence of sublethal damage to fish of various species from exposure to explosions and other less energetic sources. Fish may recover from such damage (Casper et al. 2012a, 2013b) but decreased fitness during recovery may result in their being more susceptible to predation or disease. Earlier, Gaspin (1975) observed that fish that had suffered extensive barotrauma injury from explosions generally died, although some were able to recover from extremely serious injuries if cared for in holding tanks. He concluded that fish with such injuries were unlikely to survive if released into the wild.

[1] Physostomous fishes are those species in which there is a direct connection between the swim bladder and the gut, whereas there is no such connection in physoclistous fishes. Physostomes control air in their swim bladders by either gulping air or releasing it via the gut. Physoclists have a special gland in the swim bladder wall to secrete or take up air from the swim bladder.

The physical properties of explosive shock waves that have been correlated with fish mortality are the impulse (Yelverton et al. 1975; Goertner 1978), and the maximum range in overpressure from peak positive to peak negative (Goertner 1978). The levels of particle motion are especially relevant for explosions as they may be large. However, the effects of high magnitude particle motion on fish hearing or other aspects of fish health have not been considered in any study conducted to date. It is also evident that factors like the rise time are also important (Hubbs and Rechnitzer 1952) but have not been included as variables in biological impact prediction models.

The problem for setting guidelines is that the studies that have examined the effects of explosions on fishes have each used different species, different types of explosives, and/or charges of different weights. Since the methodologies and data are so varied, the guidelines in the Tables are based on a paper representing the lowest amplitude that caused consistent mortality. Hubbs and Rechnitzer (1952) used dynamite as a source on a variety of marine species and showed a minimum amplitude of 40 – 70 psi (peak pressure) that resulted in mortality. This is the equivalent to 276 to 482 kPa, or 229 to 234 dB re 1 μPa. Other studies on rapid decompression showed that a negative pressure greater than 40% of a fish's acclimation pressure may cause barotrauma (Brown et al. 2012) severe enough to cause mortality. Simenstad (1974) determined that a 60% reduction in pressure was sufficient to rupture the swim bladder of physostomous fish. Simenstad also stressed that thresholds for damage to fish needed to be referenced to ambient pressures at the depth of exposure. For a fish acclimated to surface pressure (approximately 100 kPa), a 40% reduction in pressure would be caused by a sound wave with a negative pressure of 212 dB re 1 μPa. For each increase in acclimation depth of 10 m, the threshold in negative pressure causing significant barotrauma would increase by 6 dB re 1 μPa. The decrease in sensitivity to barotrauma with depth means that fish above an energetic source will be more susceptible to barotrauma caused by decompression, while those below the source will be less susceptible to barotrauma caused by decompression.

No data on effects of explosions on hearing (e.g., TTS) or behavior are available. However, data suggest that there may be temporary or partial loss of hearing at high sound levels, especially in fishes where the swim bladder enhances sound pressure detection. The time interval between explosions may be important when considering effects upon hearing. Rogers and Zeddies (2008) speculate that with increasing depth the gas in swim bladders will become denser, resulting in stiffening of the swim bladder and a decrease in pressure-aided hearing sensitivity. This same effect might temper the likelihood of injury to a fish's ear by higher energy sound sources even if the source were also located at depth.

There is no reason to expect masking to be of any consequence as a result of explosives. While the detection of biologically relevant sounds may be masked during an explosion, this effect would only occur during the brief duration of the sound.

Almost nothing is known about effects of explosions on the behavior of wild fishes. One likely behavioral effect is a startle reaction (e.g., a Mauthner cell mediated C-start – Eaton and Kimmel 1980; Eaton et al. 1981) if the received signal is of sufficient magnitude. Such responses last less than a second and do not necessarily result in significant changes in subsequent behavior.

Information is especially lacking on the impact of land-based explosions upon fishes. In this case, signals from seismic shots may propagate through the substrate and result in large particle motion levels being generated in adjacent shallow water bodies, including rivers and lakes. The Canadian Government has issued guidelines for the use of explosives in or near fisheries water (Wright and Hopky 1998). The guidelines state that an overpressure in excess of 100 kPa may result in adverse effects, including injury to the swim bladder, kidney, liver, spleen, and sinus venosus. Fish eggs and larvae also may be killed or damaged.

The Canadian guidelines (Wright and Hopky 1998) report that sublethal effects in fishes, including changes in behavior, have been observed on several occasions as a result of noise produced by explosives. The effects may be intensified in the presence of ice and in areas of hard substrate. The degree of effect is related to type of explosive, size, and pattern of the explosive charge(s), method of detonation, distance from the point of detonation, water depth, and species, size, and life stage of fish.

Sea Turtles: No published data are available on the specific effects of explosives on sea turtles, although preliminary studies from Ketten et al. (2005) show that fresh sea turtle cadavers are highly resistant to damage from even high level explosives. However, the explosive removal of offshore oil and gas structures in the Gulf of Mexico has resulted in the injury and death of a small number of sea turtles (Klima et al. 1988; Gitschlag and Herczeg 1994), perhaps due to the effects of rapid pressure changes on the air-filled lungs and other air-filled cavities (e.g., middle ear). As is the case for fish, recoverable damage may result in indirect mortality due to predation, increased chances of infection, changes in hormonal state, etc. Nothing is known about TTS or masking in sea turtles, nor are there any indications of behavioral changes due to exposure to explosives. However, it is expected that there are circumstances under which TTS could occur for sea turtles, as it does for all other vertebrates investigated.

Eggs and Larvae: The effects of explosions on eggs and larvae are not known, and data are needed. Damage from shock to eggs and developing embryos consist of deformation and compression of the membrane, spiral curling of the embryo, displacement of the embryo, and disruption of the vitelline membrane (Smirnov 1959). Canadian guidelines on the use of explosives near waters where there are fisheries (Wright and Hopky 1998) state that no explosive is to be detonated that is likely to produce a peak particle velocity greater than 13 mm s^{-1} in a spawning bed during the period of egg incubation.

7.5.2 Source: Pile Driving

Guidelines for pile driving are presented in Table 7.3.

Fishes: Recent studies have used a wave tube that allows controlled exposure of fish to signals replicated from actual pile driving operations (Halvorsen et al. 2011, 2012a, c; Casper et al. 2012b, 2013a, b) (see Chap. 6.2.3). Data from these studies were used to set the present guidelines for mortality. They are based on results for

Table 7.3 Pile driving. Data on mortality and recoverable injury are from Halvorsen et al. (2011, 2012a, c) based on 960 sound events at 1.2 s intervals. TTS based on Popper et al. (2005). See text for details. Note that the same peak levels are used both for mortality and recoverable injury since the same SEL_{ss} was used throughout the pile driving studies. Thus, the same peak level was derived (Halvorsen et al. 2011).

Type of Animal	Mortality and potential mortal injury	Impairment			Behavior
		Recoverable injury	TTS	Masking	
Fish: no swim bladder (particle motion detection)	>219 dB SEL_{cum} or >213 dB peak	>216 dB SEL_{cum} or >213 dB peak	>>186 dB SEL_{cum}	(N) Moderate (I) Low (F) Low	(N) High (I) Moderate (F) Low
Fish: swim bladder is not involved in hearing (particle motion detection)	210 dB SEL_{cum} or >207 dB peak	203 dB SEL_{cum} or >207 dB peak	>186 dB SEL_{cum}	(N) Moderate (I) Low (F) Low	(N) High (I) Moderate (F) Low
Fish: swim bladder involved in hearing (primarily pressure detection)	207 dB SEL_{cum} or >207 dB peak	203 dB SEL_{cum} or >207 dB peak	186 dB SEL_{cum}	(N) High (I) High (F) Moderate	(N) High (I) High (F) Moderate
Sea turtles	210 dB SEL_{cum} or >207 dB peak	(N) High (I) Low (F) Low	(N) High (I) Low (F) Low	(N) High (I) Moderate (F) Low	(N) High (I) Moderate (F) Low
Eggs and larvae	>210 dB SEL_{cum} or >207 dB peak	(N) Moderate (I) Low (F) Low	(N)Moderate (I) Low (F) Low	(N) Moderate (I) Low (F) Low	(N) Moderate (I) Low (F) Low

Notes: peak and rms sound pressure levels dB re 1 μPa; SEL dB re 1 $\mu Pa^2 \cdot s$. All criteria are presented as sound pressure even for fish without swim bladders since no data for particle motion exist. Relative risk (high, moderate, low) is given for animals at three distances from the source defined in relative terms as near (N), intermediate (I), and far (F).

Chinook salmon, Nile tilapia, hybrid striped bass, and lake sturgeon (Casper et al. 2012b, 2013a, b; Halvorsen et al. 2012c). Since these species were widely variable in body type, swim bladder configuration, and internal morphologies, the general similarity in results suggests that the guidelines may be broadly applicable to a wide range of fish species. Data are also available that show lack of effect from exposure to a SEL_{cum} of 216 dB re 1 $\mu Pa^2 \cdot s$ in hogchoker, a species without a swim bladder (Halvorsen et al. 2012c), supporting the argument that the presence of a swim bladder increases the likelihood of damage to internal body tissues.

Halvorsen et al. (2011, 2012a, c) found that the extent of injury increased with sound exposure levels and number of pile driving strikes. Thus, the guidelines here are for the lowest level where injury was found.

Guidelines for mortality and recovery for pile driving are given in terms of the dual criteria discussed in Chap. 7.1: single strike peak sound pressure level and

SEL_{cum}. The guidelines do not give the number of strikes needed to get to SEL_{cum} since this will clearly depend on the SEL_{ss} and the timing of the strikes.

Halvorsen et al. (2011, 2012a, c) demonstrated that an appropriate metric for guidelines may be a combination of the single strike SEL (SEL_{ss}) and the number of strikes that are used to yield the SEL_{cum} value, with the understanding that at the same SEL_{cum} value, higher SEL_{ss} and fewer strikes can result in the same onset of effects as a lower SEL_{ss} and more strikes.

Some data show that fish within a few meters of driving a large pile are killed (Caltrans 2004), but no data from these studies document the sound levels to which the fish were exposed or the extent of exposure before mortality occurred. At greater distances, data from caged fish show no mortality and no damage that can be clearly associated with pile driving activities (Abbott et al. 2005; Nedwell et al. 2006; Caltrans 2010a, b; Ruggerone et al. 2008; Houghton et al. 2010). Two studies looking for potential damage found no pathology in fish after exposure to pile driving (Abbott et al. 2005; Caltrans 2010a, b). However, in neither study were the fish given the opportunity or time to attain neutral buoyancy of swim bladder or tissues before exposure, and they may therefore have been protected from injury, especially the physostomous species (Stephenson et al. 2010; Halvorsen et al. 2011, 2012a, c; Casper et al. 2012b). In terms of behavioral responses, Feist (1992) and Anderson (1990) showed that fish might move away from a pile driving source.

Casper et al. (2012b, 2013a) found that fish can recover from injuries that are not mortal, although that recovery was in the laboratory under controlled conditions. In the natural environment, even these recoverable injuries could reduce fitness and lead indirectly to mortality. No data are available on TTS or masking for fish exposed to pile driving, nor are there data on behavioral responses. The TTS guidelines are based on studies using seismic airguns (Popper et al. 2005). Masking may occur for the duration that fish are exposed to pile driving but would end as soon as the pile driving ended. It is not possible to say how long behavioral effects, if any, would continue following pile driving.

Sea Turtles: Data on the effects of pile driving on sea turtles are lacking. However, Table 7.3 adopts the levels for fish that do not hear well since it is likely these would be conservative for sea turtles. Because of their rigid external anatomy, it is possible that sea turtles are highly protected from impulsive sound effects, at least with regard to pile driving and seismic airguns.

Eggs and Larvae: These Guidelines are based on recent work by Bolle et al. (2012) using a device similar to that used by Halvorsen et al. (2011, 2012a). Bolle et al. (2012) found no damage to larval fish at SEL_{cum} as high as 210 dB re 1 $\mu Pa^2 \cdot s$. Therefore the level adopted in Table 7.3 is likely to be conservative.

Eggs and larvae close to the substrate may be vulnerable to vibration associated with the ground roll generated by pile driving or seismic airguns. The Canadian guidelines for explosives referred to earlier (Wright and Hopky 1998) set a peak particle velocity of 13 mm s^{-1} for incubating eggs.

7.5.3 Source: Seismic Airguns

Guidelines for seismic airguns are found in Table 7.4.

Table 7.4 Seismic airguns. Data on mortality and recoverable injury from Halvorsen et al. (2011, 2012a, c) based on 960 sound events at 1.2 s intervals. TTS based on Popper et al. (2005). See text for details. Note that the same peak levels are used both for mortality and recoverable injury since the same SEL_{ss} was used throughout the pile driving studies. Thus, the same peak level was derived (Halvorsen et al. 2011).

Type of Animal	Mortality and potential mortal injury	Impairment			Behavior
		Recoverable injury	TTS	Masking	
Fish: no swim bladder (particle motion detection)	>219 dB SEL_{cum} or >213 dB peak	>216 dB SEL_{cum} or >213 dB peak	>>186 dB SEL_{cum}	(N) Low (I) Low (F) Low	(N) High (I) Moderate (F) Low
Fish: swim bladder is not involved in hearing (particle motion detection)	210 dB SEL_{cum} or >207 dB peak	203 dB SEL_{cum} or >207 dB peak	>>186 dB SEL_{cum}	(N) Low (I) Low (F) Low	(N) High (I) Moderate (F) Low
Fish: swim bladder involved in hearing (primarily pressure detection)	207 dB SEL_{cum} or >207 dB peak	203 dB SEL_{cum} or >207 dB peak	186 dB SEL_{cum}	(N) Low (I) Low (F) Moderate	(N) High (I) High (F) Moderate
Sea turtles	210 dB SEL_{cum} or >207 dB peak	(N) High (I) Low (F) Low	(N) High (I) Low (F) Low	(N) Low (I) Low (F) Low	(N) High (I) Moderate (F) Low
Eggs and larvae	>210 dB SEL_{cum} or >207 dB peak	(N) Moderate (I) Low (F) Low	(N) Moderate (I) Low (F) Low	(N) Low (I) Low (F) Low	(N) Moderate (I) Low (F) Low

Notes: peak and rms sound pressure levels dB re 1 μPa; SEL dB re 1 $\mu Pa^2 \cdot s$. All criteria are presented as sound pressure even for fish without swim bladders since no data for particle motion exist. Relative risk (high, moderate, low) is given for animals at three distances from the source defined in relative terms as near (N), intermediate (I), and far (F).

Fishes: The guidelines provided in Table 7.4 are derived from data from several sources. The mortality and recoverable injury guidelines for fishes and sea turtles are based on predictions derived from effects of impulsive sounds (see Chap. 7.1 for a discussion of these data), since there are no quantified data for seismic airguns. The table also adopts the dual criteria discussed for pile driving. One major difference between pile driving and seismic airguns is that it is harder to determine SEL_{cum} for airguns. This is because the received SEL_{ss} changes from shot to shot since the seismic vessel is moving and at different distances from the fish. Thus a guideline ultimately based on the closest peak level or the closest SEL_{ss} may actually be more useful than one based on the SEL_{cum}.

There are few data regarding effects of seismic airguns on fish mortality and damage to organ systems. Studies of fish with swim bladders have not shown mortality (Popper et al. 2007; Hastings et al. 2008; McCauley and Kent 2012). Data from exposure to impulsive sources suggest that airgun effects would be greater in fish with a swim bladder than in fish without a swim bladder (Halvorsen et al. 2012c; Casper et al. 2013b; and see discussion in Chap. 7.5.2 in relation to pile driving and swim bladders, and in Chap. 7.5.1 on explosives and swim bladders). Popper et al. (2005) examined the body cavity of several riverine fish species with swim bladders post-seismic exposure and found no evidence of bleeding or swim bladder damage. However, data are needed from controlled studies on more species. One study showed damage to sensory hair cells in the ear the pink snapper (*Pagrus auratus* Forster) after seismic airgun exposure (McCauley et al. 2003), while another found no damage in several other species (Popper et al. 2005; Song et al. 2008).

Guidelines for TTS are based upon data from Popper et al. (2005) for exposure of several riverine species to a seismic airgun array. Exposure to an SEL_{cum} of 186 dB re 1 $\mu Pa^2 \cdot s$ accumulated over five seismic pulses within about five minutes resulted in up to about 20 dB of TTS in the lake chub (*Couesius plumbeus*) at different frequencies, with maximum TTS at 200 and 400 Hz. The lake chub has a connection between the swim bladder and inner ear. About 20 dB of TTS occurred only at 400 Hz in adult northern pike (*Esox lucius*), a species that does not have such a connection. TTS did not occur at other frequencies, nor at any frequency in young-of-the-year northern pike. Another species without a connection between the ear and swim bladder, the broad whitefish (*Coregonus nasus*), showed no TTS to sounds after exposure at the same level (Popper et al. 2005). In all cases, fish that showed TTS recovered to normal hearing levels within 18–24 hours.

Hastings et al. (2008) examined hearing in a number of species of reef fishes exposed to a seismic survey with an airgun. The investigators found no hearing loss following sound exposures up to 190 dB re 1 $\mu Pa^2 \cdot s$ SEL_{cum} for one species in which the swim bladder is connected to the ear, and in three species where it was not. However, these data cannot be used to inform the guidelines since no information was provided about how the fish were actually exposed to the sounds.

Data on masking by seismic airgun sources are not available for any species. Masking is possible for the time that fish are exposed to the sound from airguns and may occur when animals are sufficiently far from the source for the sounds to merge and become more or less continuous (Nieukirk et al. 2004). Therefore, the guidelines show that there is a greater likelihood of masking further from the source than close to it for those fish with good hearing.

Few studies have observed the behavior of fish exposed to a seismic survey directly. Peña et al. (2013) described the real-time behavior of herring schools exposed to a full-scale 3D seismic survey, observed using sonar. No changes were observed in swimming speed, swimming direction, or school size that could be attributed to a transmitting seismic vessel as it approached from a distance of 27 km to 2 km, over a 6 h period. The unexpected lack of a response to the seismic survey was interpreted as a combination of a strong motivation for feeding by the fish, a lack of suddenness of the airgun stimulus, and an increased level of tolerance to seismic shooting.

A number of studies have demonstrated that exposure to seismic airguns has an impact on fish catch, presumably as a result of changes in fish behavior and distribution during and after sound exposure (e.g., Pearson et al. 1992; Skalski et al. 1992; Engås et al. 1996; Engås and Løkkeborg 2002; Slotte et al. 2004; Løkkeborg et al. 2012a, b). Handegard (2010) summarized the results of two surveys of the effects of seismic shooting on fish catch in Norway (Engås et al. 1996; Løkkeborg et al. 2012a, b), and compared the difference in sound exposure. In the earlier study, trawl and long-line catches of cod and haddock declined by about 50% following 5 days of shooting. In the more recent experiment, long-line catches also declined, but gill-net catches increased. The number of airgun emissions were far higher in the Løkkeborg et al. (2012a, b) experiment, but were distributed over a larger area and for a longer duration. In a central position within both areas (assuming cylindrical spreading), the total accumulated sound exposure levels (SEL_{cum}) were similar over the duration of the two experiments, but the daily levels were approximately 10 dB higher in the Engås et al. (1996) experiments. These data, however, do not inform the guidelines since the measured sound exposure levels were over long periods and received levels were not determined.

Studies on caged sandeels (*Ammodytes marinus*), a species without a swim bladder, in the North Sea, revealed distinct but minor reactions to seismic shootings (Hassel et al. 2004). No increased mortality was found during this experiment. Dalen and Knutsen (1987) observed that the distribution of a number of different species at 100 – 300 m depth changed along the course lines of a seismic airgun vessel towing an airgun array. Slotte et al. (2004) also observed that fish (Atlantic herring and blue whiting (*Micromesistius poutassou*) in an area where seismic shooting occurred, moved out of the area or to deeper waters (10–50 m deeper). Wardle et al. (2001) observed small changes in the position of pollack (*Pollachius pollachius*) in response to the firing of an airgun. However, startle responses observed from saithe (*Pollachius virens*) were elicited by the visual stimulus associated with the bubble cloud caused by airgun shots.

In conclusion, it is evident that behavioral reactions can occur to seismic airguns, but at this point there are no data that can be applied to develop guidelines.

Sea Turtles: Few data exist on the effects of seismic airgun activity on sea turtles. It is possible that seismic airgun exposure would mortally injure sea turtles that are very close to the source, although preliminary data suggest that sea turtles are highly resistant to high intensity explosives (Ketten et al. 2005), making it likely that they would also be resistant to damage from seismic airguns. It is also likely that there would be recoverable injury or TTS. Avoidance responses of sea turtles to low frequency tones have been demonstrated in caged animals (Lenhardt 1994). O'Hara and Wilcox (1990) found that sea turtles in a canal would avoid an area with an airgun, although the received level at the sea turtles was not measured. Moein et al. (1995) monitored the behavior of penned loggerhead turtles to airguns firing at 175–179 dB re 1 μPa at 1 m. Avoidance to the airguns was observed at first exposure but the sea turtles habituated to the sound over time. Behavioral responses by sea turtles, including rising to the surface and altered swimming patterns, have been elicited in caged animals exposed to an airgun at received levels of 166 dB (rms) re 1 μPa (McCauley et al. 2000). Weir (2007) reported no obvious behavioral avoidance by several species of sea turtle at the sea surface to a seismic survey as recorded by ship-based observers, although fewer turtles were seen at the surface when the airguns were firing.

Eggs and Larvae: The entries in Table 7.4 for mortality in eggs and larvae are predictions based on a recent study by Bolle et al. (2012) that indicated no damage was caused by simulated pile driving signals of 210 dB re 1 μPa^2·s SEL$_{cum}$. See Chap. 7.5.2 for a discussion of this study. Other studies suggest that eggs and larvae in very close proximity (< 5 m) to airguns are likely to suffer mortality and tissue damage (Kostyuchenko 1973; Booman et al. 1996). Sætre and Ona (1996) concluded that mortality rates caused by exposure to airgun sounds are so low compared to natural mortality that the impact from seismic surveys must be regarded as insignificant.

7.5.4 *Source: Low- and Mid-Frequency Naval Sonar*

Separate guidelines for low- and mid-frequency sonar are presented in Tables 7.5 and 7.6. The signals from military sonar have durations that are long enough to be considered continuous rather than impulsive. Therefore, guidelines are presented in terms of rms values rather than peak or SEL.

Table 7.5 Low Frequency Sonar. Masking, if it occurs, would likely only occur during the times of actual sonar transmission (see text). Data in table from Popper et al. (2007) and Kane et al. (2010).

Type of Animal	Mortality and potential mortal injury	Impairment			Behavior
		Recoverable injury	TTS	Masking	
Fish: no swim bladder (particle motion detection)	(N) Low (I) Low (F) Low	(N) Low (I) Low (F) Low	> 193 dB rms	(N) Low (I) Low (F) Low	(N) Low (I) Low (F) Low
Fish: swim bladder is not involved in hearing (particle motion detection)	>193 dB rms	> 193 dB rms	>193 dB rms	(N) Low (I) Low (F) Low	(N) Low (I) Low (F) Low
Fish: swim bladder involved in hearing (primarily pressure detection)	> 193 dB rms	> 193 dB rms	193 dB rms	(N) Moderate (I) Low (F) Low	>197 dB rms
Sea turtles	(N) Low (I) Low (F) Low	(N) Low (I) Low (F) Low	(N) Moderate (I) Moderate (F) Low	(N) Lowe (I) Low (F) Low	(N) Low (I) Low (F) Low
Eggs and larvae	(N) Low (I) Low (F) Low	(N) Low (I) Low (F) Low	(N) Low (I) Low (F) Low	(N) Low (I) Low (F) Low	(N) Low (I) Low (F) Low

Notes: rms sound pressure levels dB re 1 μPa. All criteria are presented as sound pressure even for fish without swim bladders since no data for particle motion exist. Relative risk (high, moderate, low) is given for animals at three distances from the source defined in relative terms as near (N), intermediate (I), and far (F).

Table 7.6 Mid Frequency Sonar. Data from Halvorsen et al. (2012b)

Type of Animal	Mortality and potential mortal injury	Impairment			Behavior
		Recoverable injury	TTS	Masking	
Fish: no swim bladder (particle motion detection)	(N) Low (I) Low (F) Low	(N) Low (I) Low (F) Low	NA	NA	NA
Fish: swim bladder is not involved in hearing (particle motion detection)	>210 dB rms	> 210 dB rms	NA	NA	NA
Fish: swim bladder involved in hearing (primarily pressure detection)	>210 dB rms	> 210 dB rms	>210 dB rms	(N) Low (I) Low (F) Low	> 209 dB rms
Sea turtles	NA	NA	NA	NA	NA
Eggs and larvae	NA	NA	NA	NA	NA

Notes: rms sound pressure levels dB re 1 μPa. All criteria are presented as sound pressure even for fish without swim bladders since no data for particle motion exist. Relative risk (high, moderate, low) is given for animals at three distances from the source defined in relative terms as near (N), intermediate (I), and far (F).

Fishes: Data for mortality and injury related to low-and mid-frequency sonar are based on Popper et al. (2007), Halvorsen et al. (2012b), and Kane et al. (2010), which showed no effect on the ear or non-auditory tissues when the maximum received sound pressure levels were at 193 dB re 1 μPa rms for low frequency sonar, and at 210 dB re 1 μPa rms for mid-frequency sonar. Injury, if it occurs, is thought to begin at higher sound levels than tested to date.

While studies have not been conducted with fish lacking swim bladders, it is unlikely that such fish would be damaged by sonar at the levels tested because they are not damaged by higher intensity impulsive signals (Halvorsen et al. 2012c).

Some fishes that have been exposed to low frequency sonar have shown TTS, although its duration is not yet clear (Popper et al. 2007). TTS was observed in rainbow trout (*Oncorhynchus mykiss*) and channel catfish (*Ictalurus punctatus*) exposed to LFA for several minutes at 193 dB re 1 μPa rms (Popper et al. 2007; Halvorsen et al. 2013), but not in yellow perch (*Perca flavescens*), or largemouth bass (*Micropterus salmoides*). Interestingly, not all specimens of rainbow trout showed TTS. The only TTS from mid-frequency sonar resulted from exposure at 210 dB re 1 μPa rms over five 3-s exposures in some, but not all, exposed channel catfish, a species that can detect mid-frequency sonar sounds. There was no TTS to mid-frequency sounds by rainbow trout, a species that does not hear above 1,000 Hz. The TTS in catfish of 4–6 dB recovered within 24 hours. Thus, based on the criterion that TTS must be more than 6 dB, the inclusion of this level in the guidelines Table 7.6 is highly conservative.

The only fishes in which behavior is potentially affected by mid-frequency sonar are those that have specializations that enable them to hear sounds above about 2,500 Hz (Halvorsen et al. 2012b). Higher frequency sonar (e.g., above 10 kHz) potentially affects members of the genera *Alosa* and *Brevoortia* (shads and menhadens) (Dunning et al. 1992; Nestler et al. 1992), species that can detect ultrasound (Mann et al. 1997, 2001). Behavioral responses have been shown by blueback herring (*Alosa aestivalis*) to simulated sonar pulses at frequencies from 110 to 140 kHz at sound levels above 180 dB re 1 μPa (peak) (Nestler et al. 1992).

Behavioral responses to mid-frequency naval sonar have been examined in Atlantic herring. No escape reactions (vertically or horizontally) were observed in free swimming herring exposed to sonar transmissions of 1–2 kHz and 6–7 kHz (received rms sound pressure level tested at 127–197 and 139–209 dB re 1 μPa, respectively) (Doksæter et al. 2009, 2012).

There are no data on masking by sonar. If masking occurs, it would be only be during sonar transmissions and not in the interval between transmissions. Moreover, the narrow bandwidth of most sonar would result in only a limited range of frequencies being masked.

Sea Turtles: No data are available for sea turtles and sonar. Moreover, since turtles detect sound at less than 1,000 Hz (Chap. 3.3), any effect would only be in response to low-frequency sonar.

Eggs and Larvae: No data are available on mortality or damage to eggs and larvae that could be applied to recommend guidelines. Jørgensen et al. (2005) observed increased swimming activity and startle responses in juvenile Atlantic herring exposed to sonar-like signals (1.6 and 4 kHz) in the laboratory, but no responses were observed in Atlantic cod, saithe (*Pollachius virens*), and wolf fish (*Anarhichas minor*). Some mortality occurred in clupeid eggs at the highest sound levels, but this result was not replicated. A statistical analysis showed that the likelihood of mortality to eggs and larvae from this kind of source is lower than the level of natural mortality (Kvadsheim and Sevaldsen 2005).

7.5.5 Source: Shipping and Other Continuous Noises

Recommended guidelines for shipping and other continuous sources are presented in Table 7.7.

Fishes: There is no direct evidence of mortality or potential mortal injury to fish or sea turtles from ship noise. Some evidence for auditory tissue effects or TTS caused by continuous sound comes from data on goldfish (*Carassius auratus*), an otophysan species that has specializations for enhanced sensitivity to sound pressure. Some recoverable loss of sensory hair cells occurred in the ear after 48 hours of exposure to white noise at 170 dB re 1 μPa rms (Smith et al. 2006). A maximum TTS of about 16 dB accompanied this loss. Recovery of TTS took seven days and full replacement of the sensory cells took eight days. Exposure to 158 dB re 1 μPa rms in another study also resulted in TTS in goldfish (26 dB TTS) and another pressure-sensitive fish that hears well, the catfish *Pimelodus pictus* (32 dB TTS) (Amoser and Ladich 2003). Full recovery occurred after three days for the goldfish and after fourteen days for catfish.

Table 7.7 Shipping and continuous sounds. For the most part, data in this table are based on knowing that fish will respond to sounds and their hearing sensitivity, but, as discussed in the text, there are no data on exposure or received levels that enable guideline numbers to be provided.

Type of Animal	Mortality and potential mortal injury	Impairment			Behavior
		Recoverable injury	TTS	Masking	
Fish: no swim bladder (particle motion detection)	(N) Low (I) Low (F) Low	(N) Low (I) Low (F) Low	(N) Moderate (I) Low (F) Low	(N) High (I) High (F) Moderate	(N) Moderate (I) Moderate (F) Low
Fish: swim bladder is not involved in hearing (particle motion detection)	(N) Low (I) Low (F) Low	(N) Low (I) Low (F) Low	(N) Moderate (I) Low (F) Low	(N) High (I) High (F) Moderate	(N) Moderate (I) Moderate (F) Low
Fish: swim bladder involved in hearing (primarily pressure detection)	(N) Low (I) Low (F) Low	170 dB rms for 48 h	158 dB rms for 12 h	(N) High (I) High (F) High	(N) High (I) Moderate (F) Low
Sea turtles	(N) Low (I) Low (F) Low	(N) Low (I) Low (F) Low	(N) Moderate (I) Low (F) Low	(N) High (I) High (F) Moderate	(N) High (I) Moderate (F) Low
Eggs and larvae	(N) Low (I) Low (F) Low	(N) Low (I) Low (F) Low	(N) Low (I) Low (F) Low	(N) High (I) Moderate (F) Low	(N) Moderate (I) Moderate (F) Low

Notes: rms sound pressure levels dB re 1 μPa. All criteria are presented as sound pressure even for fish without swim bladders since no data for particle motion exist. Relative risk (high, moderate, low) is given for animals at three distances from the source defined in relative terms as near (N), intermediate (I), and far (F).

Other evidence of the effects of continuous noise on hearing comes from another pressure-sensitive species, the fathead minnow (*Pimephales promelas* – Scholik and Yan 2002a), in which TTS resulted from exposure to recorded boat engine noise. However, the data for several species of fishes lacking specializations for sound pressure detection showed no TTS in response to long term noise exposure; for example, tilapia (*Oreochromis niloticus* - Smith et al. 2004b), bluegill sunfish (*Lepomis macrochirus* - Scholik and Yan 2002b), and rainbow trout (Wysocki et al. 2007). Rainbow trout exposed to increased noise (up to 150 dB re 1 μPa rms) for nine months in an aquaculture facility showed no hearing loss (as determined by AEP) nor any negative effects upon the health of the fish (Wysocki et al. 2007)

Continuous noise of any level that is detectable by fishes or sea turtles can mask signal detection, and thus may have a pervasive effect on fish behavior. However, the consequences of this masking and any attendant behavioral changes for the survival of fishes and sea turtles are unknown. Investigators have reported masked hearing thresholds for fishes exposed to noise from small boats and ferries (Scholik and Yan 2001; Vasconcelos et al. 2007). However, these studies were done using

AEP measurements, which do not reflect the ability of an animal to detect and respond to a signal against a noise background.

Ships (i.e., trawlers, ferries, research vessels) can change fish behavior (e.g., induce avoidance, alter swimming speed and direction, and alter schooling behavior) (Engås et al. 1995, 1998; Sarà et al. 2007; De Robertis and Handegard 2013). The sounds produced by motor-driven ships can cause herring to dive and swim away from the vessel (Mitson and Knudsen 2003). Paradoxically, research vessels specially designed to reduce noise can result in an even greater behavioral reaction (Ona et al. 2007). Sand et al. (2008) have pointed out that passing ships produce high levels of low-frequency noise (>10 to 1000 Hz), and that infrasonic frequencies may be responsible for the observed avoidance reactions. Several reports have attempted to define the levels at different frequencies above which fish are likely to show an avoidance response (e.g., Mitson 1995). Lack of quantification of exposure sound levels that elicit responses to ships makes it impossible to provide numerical guidelines for behavioral responses of fish to sounds from ships.

Sea Turtles: No data exist for sea turtles. Table 7.7 values are based on fish studies.

Eggs and Larvae: No data exist for eggs and larvae.

Chapter 8
Research Recommendations

In this section, recommendations are made for research that would advance the knowledge of the effects of man-made sounds on fishes and sea turtles and help the further development of exposure guidelines. A consideration of additional research questions, not necessarily directed at developing guidelines, is provided in a comprehensive gap analysis arising from a 2012 BOEM workshop (Normandeau 2012a).

8.1 General Comments

Three major issues need to be considered in planning future research on sound exposure guidelines for fishes and sea turtles. These are: (1) the importance of providing an appropriate acoustic environment for experiments; (2) the difficulties in examining behavior in response to man-made sound; and (3) dealing with the diversity of fishes.

8.1.1 The Importance of Providing an Appropriate Acoustic Environment for Experiments

Presenting carefully-measured sound stimuli to aquatic animals presents particular difficulties, especially where animals are maintained in test tanks. Major perturbations in the sound field result from wall and air interfaces surrounding test tanks (see Parvulescu 1964; Gray et al. 2015; and Rogers et al. 2015 for a discussion of this issue). As a result, much of the data on responses, behavior, and physiology from otherwise well-designed studies leave open questions on the nature of the sound field to which the animals were exposed. One major question is whether the stimulus received by the animal, and to which the animal is responding, is sound

A.N. Popper et al., *ASA S3/SC1.4 TR-2014 Sound Exposure Guidelines for Fishes and Sea Turtles*, SpringerBriefs in Oceanography, DOI 10.1007/978-3-319-06659-2_8,

pressure or particle motion. Another question is whether real sound sources can be adequately reproduced, especially when they involve transient changes in sound pressure and/or particle motion.

All fishes detect particle motion; only some species detect sound pressure (reviewed in Popper and Fay 2011). Therefore, it is critically important to monitor both the sound pressure and particle motion levels presented to fishes. Moreover, consideration should be given to the effects on the sound field resulting from the presence of the animal itself. To resolve these issues, some studies on the hearing abilities of aquatic animals have been carried out with specialized facilities designed to provide predictable and appropriate acoustic conditions. These facilities included sound ranges established in the sea (Chapman and Hawkins 1973; Schuijf 1975; Schuijf and Buwalda 1975; Popper et al. 2007; Halvorsen et al. 2012b) and specialized wave tubes (e.g., Hawkins and MacLennan 1976; Halvorsen et al. 2011, 2012a, c).

Acoustical particle motion can be measured directly with neutrally buoyant triaxial accelerometers (e.g., Zeddies et al. 2012; Martin et al. 2015) as well as computed from pressure gradient measurements using arrays of pressure-sensitive hydrophones (e.g., Popper et al. 2005; Zeddies et al. 2010; Martin et al. 2015).

8.1.2 The Difficulties in Examining Behavior

The behavioral responses of wild animals to sound are likely to vary by species, size, and age class, with animal motivation, and in different contexts. One of the fundamental truths about behavioral effects is that experiments on animals held in tanks and enclosures yield equivocal results. Captive animals do not show the wide range of behavior observed from wild animals (Oldfield 2011). They may also be damaged during capture, or their behavior may be affected by the circumstances under which they have been reared. Finally, the laboratory environment, acoustical and otherwise, is generally very different from the environment that these animals experience in the wild.

Ideally, for the purpose of establishing guidelines, behavior should be observed under conditions where the animals are well adapted to a particular location and show normal behavioral patterns. Although studies in aquaria or sea cages may provide support for larger scale field programs by providing detailed descriptions of the behavior of individuals, it is unlikely that the animals will exhibit the more complex and context-dependent behavior that they may show when not confined (Oldfield 2011).

Studying behavior in the field may be difficult and expensive, and the results are often difficult to interpret (e.g., compare Engås et al. 1996 with Løkkeborg et al. 2012a, b). However, field studies of free ranging fishes and sea turtles offer the best insights into real behavioral responses.

8.1.3 *The Diversity of Fishes and Sea Turtles*

Fishes show great diversity in their morphology, hearing physiology, and behavior. A number of species have already shown great promise as experimental subjects in hearing and sound exposure experiments, but they do not represent a wide enough range of species. Data are needed for physostomous, physoclistous, and non-swim bladder species of fish, species living at different depths, and species that have different relationships between gas bubbles and the inner ear. While initial studies on impulsive sounds suggest that the overall nature of physiological effects may be generally the same for diverse species of fishes having a swim bladder (Halvorsen et al. 2012c), there are no data on behavioral effects, although this is an area of great importance.

Sea turtles are not nearly as diverse as fishes. However, they do show significant variability in where they live, their migrations, and other aspects of their lives. Leatherbacks (Dermochelyidae) are different from the Cheloniidae. Accordingly, studies should cover representatives of both turtle families.

8.2 Research Requirements of Highest Priority

This section recommends specific high priority research areas that should be the focus for future studies on the effects of sound on fishes and sea turtles. While fishes and sea turtles are treated in separate sections, many of the same questions are relevant for both groups of animals.

8.2.1 *Fishes*

Hearing Capabilities - *Measure hearing abilities for a wider range of species, under quiet conditions, where the ratio of particle motion to sound pressure can be monitored and varied.* Studies should be done using behavioral measures wherever possible (Sisneros et al. 2015). One of the most important issues is the measurement of acoustic particle motion. No standard devices or protocols are presently available for these measurements, and these must be developed.

Hearing Loss and Mechanisms of Damage:

- *Determine the relationship between hair cell loss and hearing loss.* When loss of hair cells occurs with sound exposure, what is the change in hearing sensitivity?
- *Determine the relationship between hair cell regeneration and normal hearing ability.* What is the hearing status in fishes that have lost hair cells due to intense sound exposure, both during and after hair cell regeneration? To what extent does

normal hearing and the ability to discriminate sounds recover after exposure? Does hair cell regeneration mitigate Permanent Threshold Shift in fishes?

- *Construct biomechanical models of sound pathways to the ear and lateral line.* These models would incorporate movement of structures associated with hearing (e.g., swim bladder, otolith, hair cells) in order to predict potential effects of various sound exposures on the ear, lateral line, and swim bladder.
- TTS - *Examine the behavioral consequences of hearing loss (TTS) for the behavior and fitness of different species.* Does TTS matter, especially if it is only a few dB (e.g., less than 6 dB) and brief?
- Masking - *Examine the abilities of animals to detect, discriminate, and identify sounds with differing temporal characteristics, including natural sounds, in the presence of man-made sounds.* Masking of biologically important sounds by man-made sounds may have far reaching effects on behavior.
- *Determine the behavioral consequences of masking upon fitness using natural sounds including real-life sound sources and fish sounds.* These data will allow the effects of noise to be estimated on the range over which a signal can be detected, or the space over which sound communication occurs (acoustically active space). Studies on the ranges of acoustic communication in fishes in the presence of masking sounds are needed.
- *Determine the effect on masking of shaped sound spectra that parallel those produced by man-made sources (as opposed to white noise), and signals typical of communication sounds.* What are the general rules for predicting masking, and how does masking affect the ability to localize sound sources?
- Behavioral Effects - *Investigate the effects of various sounds and sound sources on behavior and evaluate the implications for fitness.* Because behavior is species-specific, it is difficult to generalize from one species to another. Behavioral effects will also likely change depending upon the habitat, context, animal age, time of year, and even the time of day. Studies on the behavioral effects need to be conducted in the natural environment, not in tanks.
- Injury - *Determine the characteristics of sounds that cause injury, mortality, and behavioral changes.* The development of mitigation measures depends on understanding which components of a sound produce undesirable effects.
- Cumulative Effects - *Ascertain the cumulative effects of repeated sound exposures on detection, injury, and mortality.* Such data are needed for evaluating the effects of cumulative and aggregate exposure to sound, as well as recovery from exposure. These data will also enable development of better metrics for predicting effects, comparing sources, and developing mitigation measures. How are the potentially damaging effects of intermittent sounds accumulated over time and over multiple pulses? How well do SEL and SEL_{cum} function as a metric of sound exposure to predict damage, and over what time does it make sense to measure SEL_{cum}?
- Metrics - *Determine the sound metrics that are most appropriate for predicting effects of sound exposure on fishes.* It is especially important to develop metrics based on the functional hearing groups of fishes (e.g., fishes with swim bladders mechanically linked to the ear, fishes with swim bladders, and fishes without

swim bladders). Metrics for fishes with swim bladders mechanically linked to the ear will likely be referenced to sound pressure, while those without swim bladders will likely be referenced to particle motion. It is possible that metrics for fishes with swim bladders that are not linked to the ear might be best characterized in terms of both acoustic pressure *and* acoustic particle motion. For these fishes, it will be additionally important to determine how the relative contributions of pressure and particle motion vary with sound frequency.

8.2.2 Eggs and Larvae

While most of the questions relevant to older fish are not appropriate for eggs and larvae, there are a number of questions that need to be considered, as discussed below.

- Life stages - *Examinations of effects of impulsive sounds (both pressure and particle motion) on eggs and larvae at different life stages, including physical damage and survival.*
- Settling behavior - *Closer examination of hypotheses that man-made sound can affect settlement behavior and habitat choice by larvae.*

8.2.3 Sea Turtles

Each of the topics described above for fishes applies to sea turtles. Since sea turtles species are all classified as endangered or threatened, effects of sound exposure on individual animals may be relevant. In addition to the topics above the most important sea turtle research questions are discussed below.

- Hearing - *Underwater audiograms (particularly behavioral) are needed for all species for all life stages ranging from hatchling to adult.*
- Discrimination - *How well do sea turtles discriminate between sounds that differ in frequency and/or amplitude?*
- Hearing Mechanisms - *In addition to basic audiogram data collection, studies on the mechanisms of sound transmission to the ear are needed.* Are sea turtles sensitive to acoustic pressure or particle motion? The sea turtle middle ear has air spaces at the surface that may collapse with depth. How do changes in depth affect hearing (in fish as well)?
- *Do sea turtles show TTS or PTS?* Do they have sensory hair cell damage as a result of exposure to sounds?
- Behavior - Wh*at are the behavioral implications for turtle fitness caused by exposure to man-made sound?*
- Injury - *Effects of sound on auditory and non-auditory tissues.*

Chapter 9
Summary and Conclusions

These Guidelines present the work of a Working Group established to determine broadly applicable sound exposure guidelines for fishes and sea turtles. The guidelines provided in this document are based on the best available scientific information at the time of writing. After discussion of the diversity of fish and sea turtles, guidelines were developed for broad groups of animals, defined by the way they detect sound. Animals were classified as:

- Fishes without a swim bladder (particle motion detection)
- Fishes where the swim bladder is not involved in hearing (particle motion detection)
- Fishes where the swim bladder is involved in hearing (primarily sound pressure detection)
- Fish eggs and larvae
- Sea turtles

Each sound type with a potential impact upon fishes and sea turtles was considered in terms of its acoustic characteristics and appropriate metrics defined for measurement of the received levels. The sound sources examined were:

- Explosions
- Pile driving
- Seismic airguns
- Sonar (divided into low- and mid-frequency sonar)
- Continuous sound

The potential effects of sound were divided into five types:

- Mortality and mortal injury – immediate or delayed death either due to injury or substantially reduced fitness.
- Recoverable injury – injuries, including hair cell damage, minor internal or external bleeding, etc. None of these injuries are likely to cause direct mortality.
- TTS – short or long term changes in hearing sensitivity that may, or may not, reduce fitness. TTS is defined as any change in hearing sensitivity of 6 dB or more.

A.N. Popper et al., *ASA S3/SC1.4 TR-2014 Sound Exposure Guidelines for Fishes and Sea Turtles*, SpringerBriefs in Oceanography, DOI 10.1007/978-3-319-06659-2_9,

- Masking – increase in threshold levels of detection by more than 6 dB.
- Behavioral effects – substantial change in behavior for a large portion of the animals exposed.

The proposed exposure levels for the onset of any given effect are presented as guidelines in a set of Tables. In some cases numerical guidelines are provided, expressed in appropriate metrics. When this is not possible because of lack of data, the *relative* likelihood of effects occurring was judged as "high," "moderate," or "low" for each of three distances from the source: near, medium, and far. It is important to note that the *actual* likelihood of effects will depend on the received level.

The sound exposure guidelines presented in this document should be treated as interim values, based on current information. The expectation is that with more research the guidelines can be refined. Recommendations are put forward, defining the research requirements of highest priority for extending these interim exposure guidelines.

References

Abbott R, Reyff J, Marty G (2005) Final report: monitoring the effects of conventional pile driving on three species of fish. Manson Construction Company, Richmond, CA

Amoser S, Ladich F (2003) Diversity in noise-induced temporary hearing loss in otophysine fishes. J Acoust Soc Am 113:2170–2179

Anderson JJ (1990) Assessment of the risk of pile driving to juvenile fish. In: Frauenheim JL (ed) Lessons of the 80's - Strategies of the 90's. Proceedings of the 15th Annual Member's Conference, Deep Foundations Institute, Seattle, Washington, 10–12 October 1990. Deep Foundations Institute, Hawthorne, New Jersey, p 1–11. http://www.cbr.washington.edu/papers/jim/deep.foundations.pdf. Accessed 29 Jan 2013

Andrew RK, Howe BM, Mercer, JA (2002) Ocean ambient sound: comparing the 1960s with the 1990s for a receiver off the California coast. Acoustics Research Letters Online 3:65–70

ANSI/ASA S12.64-2009/Part 1 (2009) American national standard quantities and procedures for description and measurement of underwater sound from ships – Part 1: general requirements. http://scitation.aip.org/content/asa/standard/ansi/ASASTD.ANSI.ASA.S12.64.Part.1. Accessed 14 Jan 2014

Arveson PT, Vendittis DJ (2000) Radiated noise characteristics of a modern cargo ship. J Acoust Soc Am 107:118–129

Au WWL, Hastings MC (2008) Principles of marine bioacoustics. Springer, New York

Banner A, Hyatt M (1973) Effects of noise on eggs and larvae of two estuarine fishes. Trans Am Fish Soc 1:134–136

Bartol SM, Bartol IK (2011) Hearing capabilities of loggerhead sea turtles (*Caretta caretta*) throughout ontogeny: an integrative approach involving behavioral and electrophysiological techniques. Final Report to Joint Industry Programme (JIP22 07–14)

Bartol S, Ketten DR (2006) Turtle and tuna hearing. In: Swimmer Y, Brill R (eds) Sea turtle and pelagic fish sensory biology: developing techniques to reduce sea turtle bycatch in longline fisheries. Nat Oceanic Atmos Adm Tech Memo NMFS-PIFSC-7, p 98–105

Bartol SM, Musick JA, Lenhardt M L (1999) Auditory evoked potentials of the loggerhead sea turtle *(Caretta caretta)*. Copeia 1999:836–840

Bennett DH, Falter CM, Chipps SR, Niemela K, Kinney, J (1994) Effects of underwater sound stimulating the intermediate scale measurement system on fish and zooplankton of Lake Pend Oreille, Idaho. ONR Contract N00014-92-J-4106

Brekhovskikh L, Lysanov Y (1982) Fundamentals of ocean acoustics. Springer-Verlag, New York

Bleckmann H (1993) Role of the lateral line in fish behavior. In: Pitcher TJ (ed) Fish and fisheries, Series 7, 2nd ed. Chapman and Hall, London, p 201–246

A.N. Popper et al., *ASA S3/SC1.4 TR-2014 Sound Exposure Guidelines for Fishes and Sea Turtles*, SpringerBriefs in Oceanography, DOI 10.1007/978-3-319-06659-2,

Bolle LJ, de Jong CAF, Bierman SM, van Beek PJG, van Keeken OA et al (2012) Common sole larvae survive high levels of pile-driving sound in controlled exposure experiments. PLoS One. 2012, 7(3) e33052. doi:10.1371/journal.pone.0033052

Booman C, Dalen H, Heivestad H et al (1996) Effekter av luftkanonskyting pa egg, larver og ynell (Effects of airgun shooting on eggs, larvae, and personnel). Havforskningsinstituttet, ISSN 0071–5638.

Braun CB, Grande T (2008) Evolution of peripheral mechanisms for the enhancement of sound reception. In: Webb JF, Popper AN, Fay RR (eds) Fish bioacoustics. Springer Science + Business Media, LLC. New York p 99–144

Bregman AS (1990) Auditory scene analysis: the perceptual organization of sound. MIT Press, Cambridge, MA

Brown RS, Carlson TJ, Gingerich AJ et al (2012) Quantifying mortal injury of juvenile Chinook salmon exposed to simulated hydro-turbine passage. Trans Am Fish Soc 141(1):147–157. doi: 10.1080/00028487.2011.650274

Cahn PH (ed) (1967) Lateral line detectors. Indiana University Press, Bloomington, 496 p

Caldwell J, Dragoset W (2000) A brief overview of seismic airgun arrays. The Leading Edge, August 2000:898–902

Caltrans (California Department of Transportation) (2009) Technical guidance for assessment and mitigation of the hydroacoustic effects of pile driving on fish (http://ter.ps/3xk). Appendix I, revised 2012, Compendium of pile driving sounds (http://ter.ps/3xl)

Caltrans (California Department of Transportation) (2010a) Effects of pile driving sound on juvenile steelhead. Prepared by ICF for Jones and Stokes, Seattle, WA

Caltrans (California Department of Transportation) (2010b) Necropsy and histopathology of steelhead Trout Exposed to Steel Pile Driving at the Mad River Bridges, US Highway 101, July 2009. Gary D. Marty, DVM, PhD, Fish Pathology Services, Abbotsford, BC, Canada

Caltrans (California Department of Transportation) (2004) Fisheries and hydroacoustic monitoring program compliance report for the San Francisco-Oakland Bay Bridge east span seismic safety project. Strategic Environmental Consulting, Inc. and Illingworth and Rodkin, Inc. June.

Carlson TJ (2012) Barotrauma in fish and barotrauma metrics. In: Popper AN, Hawkins AD (eds) The effects of noise on aquatic life. Springer Science + Business Media, LLC, New York, p 229–234

Carlson TJ, Johnson GE (2010) Columbia River Channel improvement project rock removal blasting: monitoring plan. Final Plan Report, Pacific Northwest National Laboratory (PNNL–19076), Richland, WA

Carlson TJ, Hastings MC, Popper AN (2007) Update on recommendations for revised sound exposure guidelines for fish during pile driving activities. CALTRANS-Arlington Memo Update 12-21-07

Carlson TJ, Johnson GE, Woodley CM, Skalski JR, Seaburg AG (2011) Compliance monitoring of underwater blasting for rock removal at Warrior Point, Columbia River Channel Improvement Project 2009/2010. Pacific Northwest National Laboratory (PNLL-20388), Richland, WA

Carlson TJ, Weiland MA (2007) Dynamic pile driving and pile driving underwater impulsive sound. Report No. WA-RD 673.1,Washington State Department of Transportation. http://worldcat.org/digitalarchive/content/cdm266301.cdmhost.com/CBT/p266401coll4/0000067163/673.1.pdf. Accessed 20 Dec 2013

Casper BM, Mann DA (2007a) Dipole hearing measurements in elasmobranchs. J Exp Biol 210:75–81

Casper BM, Mann DA (2007b) The directional hearing abilities of two species of bamboo sharks. J Exp Biol 210:505–511

Casper BM, Mann DA (2009) Field hearing measurements of the Atlantic sharpnose shark, *Rhizoprionodon terraenovae*. J Fish Biol 75:2768–2776

Casper BM, Halvorsen MB, Popper AN (2012a) Are sharks even bothered by a noisy environment? In: Popper AN, Hawkins AD (eds) The effects of noise on aquatic life. Springer Science + Business Media, New York, p 93–98

Casper BC, Popper AN, Matthews F, Carlson TJ, Halvorsen MB (2012b) Recovery of barotrauma injuries in Chinook salmon, *Oncorhynchus tshawytscha* from exposure to pile driving sound. PLoS ONE 7(6):e39593. doi:10.1371/journal.pone.0039593
Casper BM, Halvorsen MB, Matthews F, Carlson TJ, Popper AN (2013a) Recovery of barotrauma injuries resulting from exposure to pile driving sounds in two sizes of hybrid striped bass. PLoS ONE 8(9):e73844. doi:10.1371/journal.pone.0073844
Casper BM, Smith ME, Halvorsen MB et al (2013b) Effects of exposure to pile driving sounds on fish inner ear tissues. Comp Biochem Physiol A 166:352–360
Chapman CJ (1973) Field studies of hearing in teleost fish. Helgoländer wiss. Meeresunters 24:371–390
Chapman CJ, Hawkins AD (1973) A field study of hearing in the cod, *Gadus morhua* L. J Comp Physiol 85:147–167
Chapman CJ, Johnstone ADF (1974) Some auditory discrimination experiments on marine fish. J Exp Biol 61:521–528
Chapman CJ, Sand O (1974) Field studies of hearing in two species of flatfish, *Pleuronectes platessa* (L.) and *Limanda limanda* (L.) (Family Pleuronectidae). Comp Biochem Physiol 47A:371–385
Christiansen-Dalsgaard J, Brandt C, Willis KL et al (2012) Specialization for underwater hearing by the tympanic middle ear of the turtle, *Trachemys scripta elegans.* Proc Roy Soc B 279:2816–2824
Clark CW, Ellison WT, Southall BL et al (2009) Acoustic masking in marine ecosystems: intuition, analysis, and implication. Mar Ecol Prog Ser 395:201–222
Cole RH (1948) Underwater explosions. Princeton University Press, Princeton, NJ
Continental Shelf Associates, Inc. (CSA) (2004) Explosive removal of offshore structures -information synthesis report. US Department of the Interior, Minerals Management Service, Gulf of Mexico OCS Region, New Orleans, LA Study MMS 2003–070. http://www.data.boem.gov/PI/PDFImages/ESPIS/2/3042.pdf. Accessed 20 Dec 2013
Coombs S, Montgomery JC (1999) The enigmatic lateral line system. In: Fay RR, Popper AN (eds) Comparative hearing: fish and amphibians. Springer-Verlag, New York, p 319–362
Coombs S, Popper AN (1979) Hearing differences among Hawaiian squirrelfishes (Family Holocentridae) related to differences in the peripheral auditory system. J Comp Physiol 132:203–207
Coombs S, Bleckmann H, Fay RR, Popper AN (eds) (2014) The lateral line. Springer Science+Business Media, LLC, New York.
Coombs S, Görner P, Münz H (eds) (1989) The mechanosensory lateral line - neurobiology and evolution. Springer-Verlag, New York
Coppens AB, Kinsler LE, Frey AR, Sanders JV (2000) Fundamentals of acoustics, 4th edn. John Wiley and Sons, Hoboken, NJ
Corwin JT (1981) Postembryonic production and aging in inner ear hair cells in sharks. J Comp Neurol 201:541–553
Corwin, JT (1983) Postembryonic growth of the macula neglecta auditory detector in the ray, *Raja clavata*: continual increases in hair cell number, neural convergence, and physiological sensitivity. J Comp Neurol 217:345–356
Curtis KR, Howe BM, Mercer JA (1999) Low-frequency ambient sound in the North Pacific: long time series observations. J Acoust Soc Am 106:3189–3200
Dalen J, Knutsen, GM (1987) Scaring effects in fish and harmful effects on eggs, larvae and fry by offshore seismic explorations. In: Merklinger HM (ed) Progress in underwater acoustics. Plenum Press, London, p 93–102
Dekeling R, Tasker M, Ainslie M et al (2015) The European marine strategy - noise monitoring in European marine waters from 2014. Popper AN, Hawkins AD (ed) The effects of noise on aquatic life, II. Springer Science+Business Media, New York (in press)
Denton EJ, Gray, JAB (1993) Stimulation of the acoustico-lateralis system of clupeid fish by external sources and their own movements. Phil Trans Roy Soc Lond Ser B 341:113–127

Denton EJ, Gray JAB (1982) The rigidity of fish and patterns of lateral line stimulation. Nature 297:679–681

De Robertis A, Handegard NO (2013) Fish avoidance of research vessels and the efficacy of noise-reduced vessels: a review. ICES J Mar Sci 70:34–45

De Robertis A, Wilson CD, Williamson NJ (2012) Do silent ships see more fish? Comparison of a noise-reduced and a conventional research vessel in Alaska. In: Popper AN, Hawkins AD (eds) The effects of noise on aquatic life. Springer Science + Business Media, New York p 331–334

DeRuiter SL, Doukara KL (2010) Loggerhead sea turtles dive in response to airgun sound exposure. J Acoust Soc Am 127(3):1726

Dijkgraaf S (1963) The functioning and significance of the lateral-line organs. Biol Rev 38:51–105

Doksæter L, Rune G, Handegard O et al (2009) Behavioral responses of herring (*Clupea harengus*) to 1–2 and 6–7 kHz sonar signals and killer whale feeding sounds. J Acoust Soc Am 125: 554–564

Doksæter S, Handegard NO, Godø OR, Kvadsheim PH, Nordulud N (2012) Behavior of captive herring exposed to naval sonar transmissions (1.0–1.6 kHz) throughout a yearly cycle. J Acoust Soc Am 131:1623–1642

Dowding CA (2008) Construction vibrations. Prentice Hall Inc, Englewood Cliffs, NJ

Dragoset B (2000) Introduction to airguns and air-gun arrays. The Leading Edge, August, 2000:892–897

Dunning DJ, Ross QE, Geoghegan P et al 1992 Alewives in a cage avoid high-frequency sound. North Am J Fish Man 12:407–416

Durlach NI, Mason CR, Kidd G et al (2003) Note on informational masking. J Acoust Soc Am 113:2984–2987

Eaton RC, Kimmel CB (1980) Directional sensitivity of the Mauthner cell system to vibrational stimulation in zebrafish larvae. J Comp Physiol 140:337–342

Eaton RC, Lavender WA, Wieland CM (1981) Identification of Mauthner-initial-response patterns in the goldfish. Evidence from simultaneous cinematography and electrophysiology. J Comp Physiol 144:521–531

Egner SA, Mann DA (2005) Auditory sensitivity of sergeant major damselfish (*Abudefduf saxatilis*) from post-settlement juvenile to adult. Mar Ecol Prog Ser 285:213–222

Ellison ET, WT Frankel (2012) A common sense approach to source metrics. In: Popper AN, Hawkins AD (eds) The effects of noise on aquatic life. Springer Science + Business Media, LLC, New York, p 443–448

Ellison WT, Southall, BL, Clark CW, Frankel AS (2012) A new context-based approach to assess marine mammal behavioral responses to anthropogenic sounds. Cons Biol 26:21–28

Engås A, Løkkeborg S (2002) Effects of seismic shooting and vessel-generated noise on fish behaviour and catch rates. Bioacoustics 12:313–315

Engås A, Misund OA, Soldal AV, Horvei B, Solstad A (1995) Reactions of penned herring and cod to playback of original, frequency-filtered and time-smoothed vessel sound. Fisheries Res 22:243–254

Engås A, Løkkeborg S, Ona E, Soldal AV (1996) Effects of seismic shooting on local abundance and catch rates of cod (*Gadus morhua*) and haddock (*Melanogrammus aeglefinus*). Can J Fish Aquat Sci 53:2238–2249

Engås A, Haugland EK, Øvredal JT (1998) Reactions of cod (*Gadus morhua* L.) in the pre-vessel zone to an approaching trawler under different light conditions. Hydrobiologia 371/372:199–206

Fay RR (1974) Masking of tones by noise for the goldfish (*Carassius auratus*). J Comp Physiol Psychol 87:708–716

Fay RR (1988) Hearing in vertebrates: a psychophysics databook. Hill-Fay Associates, Winnetka, IL

Fay RR (1998) Auditory stream segregation in goldfish (*Carassius auratus*). Hear Res 120:69–76

Fay RR (2000) Frequency contrasts underlying auditory stream segregation in goldfish. J Assoc Res Otolaryngol 1:120–128

Fay RR (2005) Sound source localization by fishes. In: Popper AN, Fay RR (eds) Sound source localization. Springer-Verlag, New York, p 36–66

Fay RR (2008) Sound source perception and stream segregation in nonhuman vertebrate animals. In: Yost W, Popper A, Fay R (eds) Auditory perception of sound sources. Springer-Verlag, New York, p 307–323
Fay RR (2009) Soundscape and the sense of hearing in fish. Integrative Zool 4:26–32
Fay RR, Edds-Walton PL (2008) Structures and functions of the auditory nervous system of fishes. In: Webb JF, Fay RR, Popper AN (eds) Fish bioacoustics. Springer Science+Business Media, LLC, Springer, New York, p 49–97
Fay RR, Megela Simmons A (1999) The sense of hearing in fishes and amphibians. In: Fay RR, Popper AN (eds) Comparative hearing: fish and amphibians. Springer-Verlag, New York, p 269–318
Fay RR (2010) Signal-to-noise ratio for source determination and for a co-modulated masker in goldfish, *Carassius auratus*. J Acoust Soc Am, 129:3367–3372
Fay RR, Popper AN (2000) Evolution of hearing in vertebrates: the inner ears and processing. Hear Res 149:1–10
Feist BE (1992) Potential impacts of pile driving on juvenile pink (*Oncorhyncus gorbuscha*) and chub (*O. keta*) salmon behavior and distribution. MS Thesis, University of Washington
Fewtrell JL, McCauley RD (2012) Impact of airgun noise on the behaviour of marine fish and squid. Mar Pollution Bull 64:984–993
Finneran JJ, Jenkins AK (2012) Criteria and thresholds for US Navy acousic and explosive effects analysis. Report for US Navy, April 2012. http://ter.ps/3xi. Accessed 20 Dec 2013
Finneran JJ, Schlundt CE (2010) Frequency-dependent and longitudinal changes in noise-induced hearing loss in a bottlenose dolphin (*Tursiops truncatus*). J Acoust Soc Am 128:567–571
FHWG (Fisheries Hydroacoustic Working Group) (2008) Memorandum, agreement in principle for guidelines for injury to fish from pile driving activities, NOAA's Fisheries Northwest and Southwest Regions, US Fish and Wildlife Service Regions 1 and 8, California/Washington/Oregon Departments of Transportation, California Department of Fish and Game, and US Federal Highway Administration. June 12, 2008. http://www.wsdot.wa.gov/NR/rdonlyres/4019ED62-B403-489C-AF05-5F4713D663C9/0/BA_InterimCriteriaAgree.pdf. Accessed 29 Jan 2013
Fletcher LB, Crawford JD (2001) Acoustic detection by sound producing fishes: the role of gas-filled tympanic bladders. J Exp Biol 204:175–183
Friedman N (2006) The naval institute guide to world naval weapons systems, 5th edn. Naval Institute Press, Annapolis MD
Gaspin JB (1975) Experimental investigations of the effects of underwater explosions on swimbladder Fish, I: 1973 Chesapeake Bay Tests. Naval Surface Weapons Center Report NSWC/WOL/TR 75–58.
Gedamke J, Ferguson M, Harrison J et al (2015) Predicting anthropogenic noise contributions to US waters. In: Popper AN, Hawkins AD (eds) The effects of noise on aquatic life, II. Springer Science+Business Media, New York (in press)
Gitschlag GR, Herczeg BA (1994) Sea turtle observations at explosive removals of energy structures. Marine Fish Rev 56:1–8
Goertner JF (1978) Dynamical model for Explosion injury to fish. Naval Surface Weapons Center White Oak Lab, Silver Spring, MD Report No. NSWC/WOL/TR-76-155
Goertner JF, Wiley ML, Young GA, McDonald WW (1994) Effects of underwater explosions on fish without swimbladders. Naval Surface Warfare Center Report NSWC TR88-114
Govoni JJ, Settle LR, West MA (2003) Trauma to juvenile pinfish and spot inflicted by submarine detonations. J Aquatic Anim Health 15:111–119
Govoni JJ, West MA, Settle LR, Lynch RT, Greene MD (2008) Effects of underwater explosions on larval fish: implications for a coastal engineering project. J Coastal Res 24:228–233
Gray MD, Rogers PH, Popper AN, Hawkins AD, Fay RR (2015) "Large" tank acoustics: how big is big enough? In: Popper AN, Hawkins AD (eds) The effects of noise on aquatic life, II. Springer Science+Business Media, New York (in press)
Halvorsen MB, Casper BM, Woodley CM, Carlson TJ, Popper AN (2011) Predicting and mitigating hydroacoustic impacts on fish from pile installations. NCHRP Res Results Digest 363,

Project 25–28, National Cooperative Highway Research Program, Transportation Research Board, National Academy of Sciences, Washington, D.C. http://www.trb.org/Publications/Blurbs/166159.aspx. Accessed 29 Jan 2013

Halvorsen MB, Casper BM, Woodley CM, Carlson TJ, Popper AN (2012a) Threshold for onset of injury in Chinook salmon from exposure to impulsive pile driving sounds. PLoS ONE 7(6):e38968. doi:10.1371/journal.pone.0038968

Halvorsen MB, Zeddies DG, Ellison WT, Chicoine DR, Popper AN (2012b) Effects of mid-frequency active sonar on fish hearing. J Acoust Soc Am 131:599–607

Halvorsen MB, Casper BC, Matthews F, Carlson TJ, Popper AN (2012c) Effects of exposure to pile driving sounds on the lake sturgeon, Nile tilapia, and hogchoker. Proc Roy Soc B 279:4705–4714

Halvorsen MB, Zeddies DG, Chicoine D, Popper AN (2013) Effects of low frequency naval sonar exposure on three species of fish. JASA Express Letters, 134(2), EL206–210. doi:10.1121/1.4812818

Handegard N (2010) Noise dose for aquatic animals: preliminary estimates for two seismic surveys. In: Proceedings of Meetings on Acoustics 9, Acoustical Society of America, p 1–13. doi:10-1121/1.3512758

Harris GG (1964) Considerations on the physics of sound production by fishes. In: Tavolga WN (ed) Marine Bio-Acoustics. Pergamon Press, Oxford, p 233–247

Harris G, van Bergeijk WA (1962) Evidence that the lateral line organ responds to near-field displacements of sound sources in water. J Acoust Soc Am 34:1831–1841

Hassel A, Knutsen T, Dalen J et al (2004) Influence of seismic shooting on the lesser sand eel (*Ammodytes marinus*). J Marine Sci 61:1165–1173

Hastings MC (2008) Coming to terms with the effects of ocean noise on marine animals. Acoust Today 4(2):22–34

Hastings MC, Popper AN (2005) Effects of sound on fish. California Department of Transportation Contract 43A0139 Task Order 1. http://www.dot.ca.gov/hq/env/bio/files/Effects_of_Sound_on_Fish23Aug05.pdf. Accessed 20 Dec 2013

Hastings MC, Popper AN, Finneran JJ, Lanford PJ (1996) Effect of low frequency underwater sound on hair cells of the inner ear and lateral line of the teleost fish *Astronotus ocellatus*. J Acoust Soc Am 99:1759–1766

Hastings MC, Reid CA, Grebe CC, Hearn RL, Colman JG (2008) The effects of seismic airgun noise on the hearing sensitivity of tropical reef fishes at Scott Reef, Western Australia. Underwater noise measurement, impact and mitigation. Proc Inst Acoust 30(5)

Hawkins AD, Chapman CJ (1975) Masked auditory thresholds in the cod, *Gadus morhua* L. J Comp Physiol A 103:209-226

Hawkins AD, Johnstone ADF (1978) The hearing of the Atlantic salmon, *Salmo salar*. J Fish Biol 13:655–673

Hawkins AD, MacLennan DN (1976) An acoustic tank for hearing studies on fish. In: Schuijf A, Hawkins AD (eds) Sound reception in fish. Elsevier, The Netherlands, p 149–169

Hawkins AD, Sand O (1977) Directional hearing in the median vertical plane by the cod. J Comp Physiol 122:1–8

Hawkins AD, Popper AN, Wahlberg M (eds) (2008) International conference on the effects of noise on aquatic life. Bioacoustics 17:1–350

Hazelwood RA (2012) Ground roll waves as a potential influence on fish: measurement and analysis techniques. In: Popper AN, Hawkins AD (eds) The effects of noise on aquatic life. Springer Science + Business Media, New York, p 449–452

Hazelwood RA, Macey PC (2015) The intrinsic directional information of ground roll waves. Popper AN, Hawkins AD (eds) The effects of noise on aquatic life, II, Springer Science+Business Media, New York (in press)

Helfman GS, Collette BB, Facey DE, Bowen BW (2009). The diversity of fishes: biology, evolution, and ecology. Blackwell Science, Malden, MA

Henderson D, Hamernik RP (2012) The use of kurtosis measurement in the assessment of potential noise trauma. In: Le Prell CG, Henderson D, Fay RR, Popper AN (eds) Noise induced hearing loss. Springer Science+Business Media, LLC, New York, p 41–56

Higgs DM, Souza MJ, Wilkins HR, Presson JC, Popper AN (2002) Age- and size-related changes in the inner ear and hearing ability of the adult zebrafish (*Danio rerio*). J Assn Res Otolaryngol 3:174–184
Hildebrand JA (2009) Anthropogenic and natural sources of ambient noise in the ocean. Mar Ecol Prog Ser 395:5–20
Houghton J, Starkes J, Stutes J et al (2010) Acoustic monitoring of *in situ* exposures of juvenile coho salmon to pile driving noise at the port of Anchorage Marine Terminal redevelopment project, Knik Arm, Alaska. Paper presented at Alaska Marine Sciences Symposium, Anchorage, January 2010
Hovem JM, Tronstad TV, Karlsen HE, Løkkeborg S (2012) Modeling propagation of seismic airgun sounds and the effects on fish behavior. IEEE J Oceanic Eng 37:576–588
Hubbs CL, Rechnitzer AB (1952) Report on experiments designed to determine effects of underwater explosions on fish life. Calif Fish Game 38:333–366
Illingsworth R, Rodkin R (2001) Noise and vibration measurements associated with the pile installation demonstration project for the San Francisco-Oakland Bay Bridge east span. Technical Report of Illingworth & Rodkin, Inc. Petaluma, CA
Illingsworth R, Rodkin R (2007) Compendium of pile driving sound data. Prepared for the California Department of Transportation, Sacramento, CA, September 27, 2007
ISO/DIS 1683 (2013) Acoustics – Preferred reference values for acoustical and vibratory levels. http://www.iso.org/iso/home/store/catalogue_tc/catalogue_detail.htm?csnumber=64648. Accessed 31 Jan 2014
ISO 80000–8:2007 (2007) Quantities and units Part 8: Acoustics. http://www.iso.org/iso/catalogue_detail?csnumber=40669. Accessed 16 Jan 2014
ISO/PAS 17208–1:2012 (2012) Acoustics - quantities and procedures for description and measurement of underwater sound from ships - Part 1: General requirements for measurements in deep water. http://scitation.aip.org/search?value1=17208-1%3A2012&option1=all&option12=resultCategory&value12=ResearchPublicationContent. Accessed 14 Jan 2014
Jerkø H, Turunen-Rise I, Enger PS, Sand O (1989) Hearing in the eel (*Anguilla anguilla*). J Comp Physiol A 165:455–459
Johnson CE (2012) Regulatory assessments of the effects of noise: moving from threshold shift and injury to behavior. In: Popper AN, Hawkins AD (eds) The effects of noise on aquatic life. Springer Science + Business Media, LLC, New York, p 563–565
Jørgensen R, Olsen KK, Falk-Petersen IB, Kanapthippilai P (2005) Investigations of potential effects of low frequency sonar signals on survival, development and behaviour of fish larvae and juveniles. Norwegian College of Fishery Science, University of Tromsø Norway
Kalmijn AJ (1988) Hydrodynamic and acoustic field detection. In: Atema, J, Fay RR, Popper AN, Tavolga WN (eds) Sensory biology of aquatic animals. Springer-Verlag, New York, p 83–130
Kane AS, Song J, Halvorsen MB et al (2010) Exposure of fish to high intensity sonar does not induce acute pathology. J Fish Biol 76:1825–1840
Keevin TM, Hempen GL (1997) The environmental effects of underwater explosions with methods to mitigate impacts. US Army Corps of Engineers, St Louis District, St. Louis, Mo
Ketten DR, Cramer S, Arruda J et al (2005) Experimental measures of blast trauma in sea turtles. Symposium on Environmental Consequences of Underwater Sound, Office of Naval Research, Arlington, VA
Klima EF, Gitschlag GR, Renaud ML (1988) Impacts of the explosive removal of offshore petroleum platforms on sea turtles and dolphins. Marine Fisheries Rev 50:33–42
Knight CR, Swaddle JP (2011) How and why environmental noise impacts animals: an integrative, mechanistic review. Ecol Lett. doi:10.1111/j.1461-0248.2011.01664.x
Kostyuchenko LP (1973) Effects of elastic waves generated in marine seismic prospecting on fish eggs in the Black Sea. Hydrobiol J 9:45–46
Kvadsheim PH, Sevaldsen EM (2005) The potential impact of 1–8 kHz active sonar on stocks of juvenile fish during sonar exercises. FFI/Report-2005/01027.
Ladich F, Fay RR (2013) Auditory evoked potential audiometry in fish. Rev Fish Biol Fisheries. doi:10.1007/s11160-012-9297-z

Laughlin, J (2006) Underwater sound levels associated with pile driving at the Cape Disappointment boat launch facility, wave barrier project. Washington State Parks Wave Barrier Project Underwater Technical Report, 13 March 2006

Lavender AL, Bartol SM, Bartol IK (2012) Hearing capabilities of loggerhead sea turtles (*Caretta caretta*) throughout ontogeny. In: Popper AN, Hawkins AD, (eds) The effects of noise on aquatic life. Springer Science + Business Media, LLC, New York, p 89–92

Laws, R. (2012) Cetacean hearing-damage zones around a seismic source. In: Popper AN, Hawkins AD (eds) The effects of noise on aquatic life. Springer Science + Business Media, LLC, New York, p 473–476

Laws RH, Hedgeland D (2008) The marine seismic airgun. Bioacoustics 17:124–126.

Le Prell CG, Henderson D, Fay RR, Popper AN (eds). (2012) Noise-induced hearing loss: from animal models to human trials. Popper AN, Hawkins AD (eds) The effects of noise on aquatic life. Springer Science + Business Media, LLC, New York

Lenhardt ML (1994) Seismic and very low frequency sound induced behaviors in captive loggerhead marine sea turtles (*Caretta caretta*). In: Bjorndal KA, Bolten AB, Johnson DA, Eliazar PJ (eds) Proceedings of the Fourteenth Annual Symposium on Sea Turtle Biology and Conservation, Springfield, VA, NOAA Technical Memorandum NMFS-SEFSC-351, p 238–241

Lenhardt ML, Klinger RC, Musick JA (1985) Marine turtle middle-ear anatomy. J Aud Res 25:66–72

Lewandowski J, Burkhard E, Skrupky K, Epperson D (2012) United States Bureau of Ocean Energy Management, Regulation and Enforcement: Filling data gaps to better understand the effects of anthropogenic noise on marine life. Popper AN, Hawkins AD (eds) The effects of noise on aquatic life. Springer Science + Business Media, LLC, New York, p 567–570

Liberman MC (2015) Noise-induced hearing loss: permanent vs. temporary threshold shifts and the effects of hair-cell vs. neuronal degeneration. In: Popper AN, Hawkins AD (eds) The effects of noise on aquatic life, II. Springer Science+Business Media, New York (in press)

Løkkeborg S, Soldal AV (1993) The influence of seismic exploration with airguns on cod (*Gadus morhua*) behaviour and catch rates. ICES Marine Sci Symp 196:62–67

Løkkeborg S, Ona E, Soldal A, Salthaug A (2012a) Effects of sounds from seismic airguns on fish behavior and catch rates. In: Popper AN, Hawkins AD (eds) The effects of noise on aquatic life. Springer Science + Business Media, New York, p 415–419

Løkkeborg S, Ona E, Vold A, Salthaug A (2012b) Sounds from seismic airguns: gear- and species-specific effects on catch rates and fish distribution. Can J Fish Aquat Sci 69:1278–1291

Lombarte A, Popper AN (1994) Quantitative analyses of postembryonic hair cell addition in the otolithic endorgans of the inner ear of the European hake, *Merluccius merluccius* (Gadiformes, Teleostei). J Comp Neurol 345:419–428

Lombarte A, Yan HY, Popper AN, Chang JC, Platt C (1993) Damage and regeneration of hair cell ciliary bundles in a fish ear following treatment with gentamicin. Hear Res 66:166–174

Maes J, Turnpenny AWH, Lambert, DR et al (2004) Field evaluation of a sound system to reduce estuarine fish intake rates at a power plant cooling water inlet. J Fish Biol 64:938–946

MacGillivray AO, Chapman RN (2005) Results from an Acoustic Modeling Study of Seismic Airgun Survey Noise in Queen Charlotte Basin. University of Victoria Report, December 7, 2005

Mann DA, Higgs DM, Tavolga WN, Souza MJ, Popper AN (2001) Ultrasound detection by clupeiform fishes. J Acoust Soc Am 109:3048–3054

Mann DA, Lu Z, Popper AN (1997) A clupeid fish can detect ultrasound. Nature 389: 341

Martin B, Zeddies DG, Gaudet B (2015) Evaluation of three sensor types for particle motion measurement. In: Popper AN, Hawkins AD (eds) The effects of noise on aquatic life, II. Springer Science+Business Media, New York (in press)

Martin KJ, Alessi SC, Gaspard JC et al (2012) Underwater hearing in the loggerhead turtle (*Caretta caretta*): a comparison of behavioral and auditory evoked potential audiograms. J Exp Biol 215:3001–3005

Mattsson A, Parkes G, Hedgeland D (2012) Svein Vaage Broadband Airgun Study. In: Popper AN, Hawkins AD (eds) The effects of noise on aquatic life. Springer Science + Business Media, LLC, New York, p 473–475

McCauley RD, Kent CS (2012) A lack of correlation between airgun signal pressure waveforms and fish hearing damage. In: Popper AN, Hawkins AD (eds) The effects of noise on aquatic life. Springer Science+Business Media, LLC, New York, p 245–250

McCauley RD, Fewtrell J, Duncan AJ et al (2000) Marine seismic surveys – a study of environmental implications. APPEA J 40:692–706

McCauley RD, Fewtrell J, Popper AN (2003) High intensity anthropogenic sound damages fish ears. J Acoust Soc Am 113:638–642

McDonald MA, Hildebrand JA, Wiggins SM (2006) Increases in deep ocean ambient noise in the Northeast Pacific west of San Nicolas Island, California. J Acoust Soc Am 120:711–718

McKenna MF, Ross D, Wiggins SM, Hildebrand JA (2012) Underwater radiated noise from modern commercial ships. J Acoust Soc Am 131:92–103

McKinstry C, Carlson T, Brown R (2007) Derivation of a mortal injury metric for studies of rapid decompression of depth-acclimated physostomous fish. PNNL-17080, Pacific Northwest National Laboratory, Richland, WA. http://www.pnl.gov/main/publications/external/technical_reports/PNNL-17080.pdf. Accessed 20 Dec 2013

Mitson RB (1995) Underwater noise of research vessels. Cooperative Research Report, International Council for the Exploration of the Sea. p 1–65. http://www.brasil-rounds.gov.br/round9/round9/guias_R9/sismica_R9/Bibliografia/Mitson%201995%20-%20Underwater%20noise%20of%20research%20vessels.pdf. Accessed 20 Dec 2013

Mitson RB, Knudsen HP (2003) Causes and effects of underwater noise on fish abundance estimation. Aquatic Liv Res 16:255–263

Moein SE, Musick JA, Keinath JA et al (1995) Evaluation of seismic sources for repelling sea turtles from hopper dredges. In: Hales LZ (ed) Sea turtle research program: summary report. Prepared for US Army Engineer Division, South Atlantic, Atlanta, GA, and US Naval Submarine Base, Kings Bay, GA. Technical Report CERC-95, 90, p 75–78

Montgomery J, Coombs, S (1996) Biology of the mechanosensory lateral line in fishes. Rev Fish Biol Fisheries 5:399–416

Montgomery J, Bleckmann H, Coombs, S (2014) Sensory ecology and neuroethology of the lateral line. In: Coombs S, Bleckmann H, Fay RR, Popper AN (eds) The lateral line. Springer Science+Business Media, LLC, New York, p 121–150

Mueller C, McGregor PK, Gill AB et al (2010) Effects of pile-driving noise on the behaviour of marine fish. Cowrie Technical report, Fish 06–08 / Cefas Ref: C3371

Myrberg AA Jr, Spires JY (1980) Hearing in damselfishes: an analysis of signal detection among closely related species. J Comp Physiol 140:135–144

Nedwell JR, Turnpenny AWH, Lovell JM, Edwards B (2006) An investigation into the effects of underwater piling noise on salmonids. J Acoust Soc Am 120:2550–2554

Nedwell JR, Turnpenny AWH, Lovell J et al (2007) A validation of the dB_{ht} as a measure of the behavioural and auditory effects of underwater noise. Subacoustech Report No 534R1231

Nestler JM, Ploskey GR, Pickens J, Menezes J, Schilt C (1992) Responses of blueback herring to high-frequency sound and implications for reducing entrainment at hydropower dams. N A J Fish Manage 12:667–683

Nieukirk SL, Stafford KM, Mellinger DK, Dziak RP, Fox CG (2004) Low-frequency whale and seismic airgun sounds recorded in the mid-Atlantic Ocean. J Acoust Soc Am 115: 1832–1843

Normandeau Associates, Inc (2012a) Effects of noise on fish, fisheries, and invertebrates in the US Atlantic and Arctic from energy industry sound-generating activities. A Workshop Report for the US Dept of the Interior, Bureau of Ocean Energy Management. Contract # M11PC00031.

Normandeau Associates, Inc (2012b) Effects of noise on fish, fisheries, and invertebrates in the U.S. Atlantic and Arctic from energy industry sound-generating activities. A Literature Synthesis for the US Dept of the Interior, Bureau of Ocean Energy Management. Contract # M11PC00031

NRC (National Research Council) (2003) Ocean Noise and Marine Mammals. National Academy Press, Washington, DC

NRC (National Research Council) (2005) Marine mammal populations and ocean noise: determining when noise causes biologically significant effects. National Academy Press, Washington, DC
Nyquist H (2002) Certain topics in telegraphic transmission theory. Proceedings of the IEEE 90:280–305. http://ieeexplore.ieee.org/xpl/login.jsp?tp=&arnumber=989875&url=http%3A%2F%2Fieeexplore.ieee.org%2Fxpls%2Fabs_all.jsp%3Farnumber%3D989875. Accessed 20 Dec 2013
O'Hara J, Wilcox JR (1990) Responses of loggerhead sea turtles, *Caretta caretta*, to low frequency sound. Copeia 199:564–567
Oldfield RG (2011) Aggression and welfare in a common aquarium fish, the Midas cichlid. J of Applied Animal Welfare Science, 14(4):340–360. doi:10.1080/10888705.2011.600664
International Association of Oil & Gas Producers (OGP) (2011) Model based assessment of underwater noise from an airgun array soft start operation. Report 451, Feb 2011. http://www.ogp.org.uk/pubs/451.pdf. Accessed 20 Dec 2013
Ona E, Godø OR, Handegard NO et al (2007) Silent research vessels are not quiet. J Acoust Soc Am 121:EL145–EL150
Parvulescu A (1964) Problems of propagation and processing. In: Tavolga WN (ed) Marine Bioacoustics. Pergamon Press, Oxford, p 87–100
Pearson WH, Skalski JR, Malme CI (1992) Effects of sounds from a geophysical survey device on behavior of captive rockfish (*Sebastes* ssp). Can J Fish Aquat Sci 49: 1343–1356.
Peña H, Handegard NO, Ona E (2013) Feeding herring schools do not react to seismic airgun surveys. ICES J of Marine Science. doi:10.1093/icesjms/fst079
Piniak WED, Mann DA, Eckert SA, Harms CA (2012) Amphibious Hearing in Sea turtles. In: Popper AN, Hawkins AD (eds) The effects of noise on aquatic life. Springer Science + Business Media, LLC, New York, p 83–88
Popper AN (1972) Pure-tone auditory thresholds for the carp *Cyprinus carpio*. J Acoust Soc Am 52:1714–1717
Popper AN, Clarke NL (1976) The auditory system of the goldfish (*Carassius auratus*): effects of intense acoustic stimulation. Comp Biochem Physiol A 53:11–18
Popper AN, Coombs S (1982) The morphology and evolution of the ear in Actinopterygian fishes. Amer Zool 22:311–328
Popper AN, Fay RR (2011) Rethinking sound detection by fishes. Hear Res 273:25–36
Popper AN, Hastings MC (2009) Effects of anthropogenic sources of sound on fishes. J Fish Biol. 75:455–498
Popper AN, Hawkins AD (eds) (2012) The effects of noise on aquatic life. Springer Science+Business Media, LLC, New York
Popper AN, Hawkins AD (eds) (2015) The effects of noise on aquatic life, II. Springer Science+Business Media, LLC, New York
Popper AN, Hoxter B (1984) Growth of a fish ear: I. Quantitative analysis of sensory hair cell and ganglion cell proliferation. Hear Res 15:133–142
Popper AN, Schilt CR (2008) Hearing and acoustic behavior (basic and applied). In: Webb JF, Fay RR, Popper AN (eds). Fish bioacoustics. Springer Science+Business Media, LLC, New York, p 17–48
Popper AN, Tavolga WN (1981) Structure and function of the ear of the marine catfish, *Arius felis*. J Comp Physiol 144:27–34
Popper AN, Fay RR, Platt C, Sand O (2003) Sound detection mechanisms and capabilities of teleost fishes. In: Collin SP, Marshall NJ (eds) Sensory processing in aquatic environments. Springer-Verlag, New York, p 3–38
Popper AN, Smith ME, Cott PA et al (2005) Effects of exposure to seismic airgun use on hearing of three fish species. J Acoust Soc Am 117:3958–3971
Popper AN, Halvorsen MB, Kane E et al (2007) The effects of high-intensity, low-frequency active sonar on rainbow trout. J Acoust Soc Am 122:623–635
Reyff JA (2008) Underwater sound pressure levels associated with marine pile driving: assessment of impacts and evaluation of control measures. J Transportation Res Board CD 11-S:481–490
Reyff JA (2012) Underwater sounds from unattenuated and attenuated marine pile driving. In: Popper AN, Hawkins AD (eds) The effects of noise on aquatic life. Springer Science + Business Media, LLC New York, p 439–444

Richardson WJ, Greene CR Jr, Malme CI, Thomson DH (1995) Marine mammals and noise. Academic Press, New York, 577 p

Ridgway SH, Wever EG, McCormick JG, Palin J, Anderson JH (1969) Hearing in the giant sea turtle, *Chelonia mydas*. Proc Nat Acad Sci USA 64(3):884–890

Robinson SP, Theobald PD, Hayman G et al (2011) Measurement of underwater noise from marine aggregate dredging operations. Final Report. MALSF Ref No: MEPF 09/p108

Rogers PH, Cox M (1988) Underwater sound as a biological stimulus. In: Atema J, Fay RR, Popper AN, WN Tavolga (eds) Sensory biology of aquatic animals. Springer-Verlag, New York, p 131–149

Rogers PH, Zeddies DG (2008) Multipole mechanisms for directional hearing in fish. In: Webb JF, Popper AN, Fay RR (eds) Fish bioacoustics. Springer-Verlag, New York, p 233–252

Rogers PH, Hawkins AD, Popper AN, Fay RR, Gray MD (2015) Parvulescu revisited: small tank acoustics for bioacousticians. In: Popper AN, Hawkins AD (eds) The effects of noise on aquatic life, II. Springer Science+Business Media, New York (in press)

Ross D (1987) Mechanics of Underwater Noise. Península Publishing, Los Altos, CA

Ross D (1993) On ocean underwater ambient noise. Acoustics Bull 18:5–8

Ruggerone GT, Goodman SE, Miner R (2008) Behavioral response and survival of juvenile coho salmon to pile driving sounds. Natural Resources Consultants, Inc. for Port of Washington, 42p

Sætre R, Ona E (1996) Seismiske undersøkelser og skader på fiskeegg og -larver; en vurdering av mulige effekter på bestandsnivå. Havforskningsinstituttet, Fisken og Havet nr. 8–1996. Seismic investigations and damage to fish eggs and larvae: an assessment of potential effects on the population level.

Sand O, Bleckmann H (2008) Orientation to auditory and lateral line stimuli. In: Webb JF, Popper AN, Fay RR (eds) Fish bioacoustics. Springer, New York, p 183–231

Sand O, Enger PS (1973a) Evidence for an auditory function of the swimbladder in the cod. J Exp Biol 59:405–414

Sand O, Enger PS (1973b) Function of the swimblabber in fish hearing. In: Moller A (ed) Basic mechanisms of hearing. Academic Press, New York, p 893–908

Sand O, Hawkins AD (1973) Acoustic properties of the cod swim bladder. J Exp Biol 58: 797–820

Sand O, Hawkins AD (1974) Measurements of swim bladder volume and pressure in the cod. Norw J Zool 22:31–34

Sand O, Karlsen HE, Knudsen FR (2008) Comment on "Silent research vessels are not quiet." J Acoust Soc Am 123:1831–1833

Sarà G, Dean JM, D'Amato D et al (2007) Effect of boat noise on the behaviour of bluefin tuna *Thunnus thynnus* in the Mediterranean Sea. Marine Ecology Prog Ser 331:243–253

Scholik AR, Yan HY (2001) Effects of underwater noise on auditory sensitivity of a cyprinid fish. Hear Res 152:17–24

Scholik AR, Yan HYY (2002a) Effects of boat engine noise on the auditory sensitivity of the fathead minnow *Pimephales promelas*. Environ Biol Fishes 63:203–209.

Scholik AR, Yan HY (2002b) The effects of noise on the auditory sensitivity of the bluegill sunfish, *Lepomis macrochirus*. Comp Biochem Physiol A 133:43–52

Schuck JB, Smith ME (2009) Cell proliferation follows acoustically-induced hair cell bundle loss in the zebrafish saccule. Hear Res 253:67–76

Schuijf A (1975) Directional hearing of cod (*Gadus morhua*) under approximate free field conditions. J Comp Physiol A 98:307–332

Schuijf A, Buwalda RJA (1975). On the mechanism of directional hearing in cod (*Gadus morhua* L). J Comp Physiol 98:333–343

Schuijf A, Hawkins AD (1983) Acoustic distance discrimination by the cod. Nature 302: 143–144

Siler W (1969) Near- and far-fields in a marine environment. J Acoust Soc Am 46:483–484

Simenstad CA (1974) Biological effects of underground nuclear testing on marine organisms. I. Review of documented shock effects, discussion of mechanisms of damage, and prediction of Amchitka test effects. In: Young GA (compiler), Proceedings of the First Conference on the

Environmental Effects of Explosive and Explosions (30–31 May 1973). NOLTR 73–223, Naval Ordinance Laboratory, Silver Spring, Maryland. p. 86–97
Sisneros JA, Popper AN, Hawkins AD, Fay RR (2015) Auditory Evoked Potential audiograms compared to behavioral audiograms in aquatic animals. In: Popper AN, Hawkins AD (eds) The effects of noise on aquatic life, II. Springer Science+Business Media, New York (in press)
Skalski JR, Pearson WH, Malme CI (1992) Effects of sounds from a geophysical survey device on catch-per-unit-effort in a hook-and-line fishery for rockfish (*Sebastes* ssp.). Can J Fish Aquat Sci 49:1357–1365
Slabbekoorn H, Bouton N (2008) Soundscape orientation: a new field in need of sound investigation. Anim Behav. doi:0.1016/j.anbehav.2008.06.010
Slabbekoorn H, Bouton N, van Opzeeland I et al (2010) A noisy spring: the impact of globally rising underwater sound levels on fish. Trends Ecol Evol 25:419–427
Slotte A, Kansen K, Dalen J, Ona E (2004) Acoustic mapping of pelagic fish distribution and abundance in relation to a seismic shooting area off the Norwegian west coast. Fish Res 67:143–150
Smirnov AI (1959) The effect of mechanical agitation in different periods of development on the eggs of autumn chum salmon (*Oncorhyncus keta* INFRASP *Autumnalis* Berg, Salmonidae). Fish Res Board Canada Translation Ser 20 (originally in Russian). Doklady Akademii Nauk SSSR 105, 873–876 (1955)
Smith ME (2012) Predicting hearing loss in fishes. In: Popper AN, Hawkins AD (eds) The effects of noise on aquatic life. Springer Science + Business Media, LLC, New York, p 571–574
Smith ME (2015) The relationship between hair cell loss and hearing in fishes. In: Popper AN, Hawkins AD (eds) The effects of noise on aquatic life, II. Springer Science+Business Media, New York (in press)
Smith ME, Coffin AB, Miller DL, Popper AN (2006) Anatomical and functional recovery of the goldfish (*Carassius auratus*) ear following noise exposure. J Exp Biol 209:4193–4202
Smith ME, Kane AS, Popper AN (2004a) Noise-induced stress response and hearing loss in goldfish (*Carassius auratus*). J Exp Biol 207:427–435
Smith ME, Kane AS, Popper AN (2004b) Acoustical stress and hearing sensitivity in fishes: does the linear threshold shift hypothesis hold water? J Exp Biol 207:3591–3602
Smith ME, Schuck JB, Gilley RR, Rogers BD (2011) Structural and functional effects of acoustic exposure in goldfish: evidence for tonotopy in the teleost saccule. BMC Neurosci 12:19. doi:10.1186/1471-2202-12-19
Song J, Mann DA, Cott PA, Hanna BW, Popper AN (2008) The inner ears of northern Canadian freshwater fishes following exposure to seismic airgun sounds. J Acoust Soc Am 124:1360–1366
Southall BL (2005) Final report of the 2004 International symposium shipping noise and marine mammals: a forum for science, technology, and management. National Marine Fisheries Service, Office of Protected Resources, Technical Report. National Oceanic and Atmospheric Administration, Washington, DC. http://www.beamreach.org/wiki/images/4/47/2004NoiseReport.pdf. Accessed 29 Jan 2013
Southall BL, Bowles AE, Ellison WT et al (2007) Marine mammal noise exposure guidelines: initial scientific recommendations. Aquatic Mamm 33:411–521
Stadler JH, Woodbury DP (2009) Assessing the effects to fishes from pile driving: application of new hydroacoustic guidelines. Inter-Noise 2009
Stephenson JR, Gingerich AJ, Brown RS et al (2010) Assessing barotrauma in neutrally and negatively buoyant juvenile salmonids exposed to simulated hydro-turbine passage using a mobile aquatic barotrauma laboratory. Fisheries Res, 106:271–278
Tashmukhambetov AM, Sidorovskaia NA, Ioup GE, Ioup JW (2008) Modeling the three-dimensional field of a seismic airgun array and comparison to 2003 measured data. J Acoust Soc Am 124:2599–2599
Tasker M (2012) Regulation of sound in the ocean: recent and future possible changes. In: Popper AN, Hawkins AD (eds) The effects of noise on aquatic life. Springer Science + Business Media, LLC, New York, p 571–574

Tasker ML (2015) How might we assess and manage the effects of underwater noise on populations of marine animals? In: Popper AN, Hawkins AD (eds) The effects of noise on aquatic life, II. Springer Science+Business Media, New York (in press)
Tavolga WN (1967) Masked auditory thresholds in teleost fishes. In: Tavolga WN (ed) Marine Bio-Acoustics, II. Pergamon Press, Oxford, p 233–245
Tavolga WN (1974) Signal/noise ratio and the critical band in fishes. J Acoust Soc Am 55: 1323–1333
TNO (Nederlandse Organisatie voor Toegepast Natuurwetenschappelijk Onderzoek) (2011) Standard for measurement and monitoring of underwater noise, Part I: physical quantities and their units. TNO report: TNO-DV 2011 C235
Urick RJ (1983). Principles of underwater sound. McGraw-Hill Book Company, New York
Vabø R, Geir H, Anders F, Jørgensen T, Løkkeborg S, Skret G (2004). Simulating search behaviour of fish towards bait. ICES J Mar Sci 61:1224–1232
van Bergeijk WA (1964) Directional and nondirectional hearing in fish. In: Tavolga WN (ed) Marine bio-acoustics. Pergamon Press, Oxford, p 281–299
van Bergeijk WA (1967). The evolution of vertebrate hearing, In: Neff WD (ed) Contributions to physiology. Academic Press, New York, p 1–49
Vasconcelos RO, Amorim MCP, Ladich F (2007) Effects of ship noise on the detectability of communication signals in the Lusitanian toadfish. J Exp Biol 210: 2104–2112
Wardle CS, Carter TJ, Urquhart GG et al (2001) Effects of seismic airguns on marine fish. Cont Shelf Res 21: 1005–1027
Webb JF, Fay RR, Popper AN (eds) (2008). Fish bioacoustics. Springer Science+Business Media, LLC, New York
Weir CR (2007) Observations of marine sea turtles in relation to seismic airgun sound off Angola. Mar Turtle Newsletter 116:17–20
Wever EG, Vernon JA (1956) Sound transmission the turtle's ear. Proc Nat Acad Sci 42: 229–299
Wever EG (1978) The Reptile Ear. Princeton University Press, Princeton, NJ
Willis KL (2015) Underwater hearing in turtles. In: Popper AN, Hawkins AD (eds) The effects of noise on aquatic life, II. Springer Science+Business Media, New York (in press)
Woodbury D, Stadler J (2008) A proposed method to assess physical injury to fishes from underwater sound produced during pile driving. Bioacoustics 17:289–291
Wright DG, Hopky GE 1998 Guidelines for the use of explosives in or near Canadian fisheries waters. Can Tech Rep Fis Aquat Sci 2107: iv + 34p
Wright KJ, Higgs DM, Leis JM (2011) Ontogenetic and interspecific variation in hearing ability in marine fish larvae. Mar Ecol Prog Ser 424:1–13
Wysocki LE, Davidson JW III, Smith ME et al (2007) Effects of aquaculture production noise on hearing, growth, and disease resistance of rainbow trout *Oncorhynchus mykiss*. Aquaculture 272:687–697
Yelverton JT, Richmond DR, Hicks W, Saunders K, Fletcher ER (1975) The relationship between fish size and their response to underwater blast. Report DNA 3677T, Director, Defense Nuclear Agency, Washington, DC
Zeddies DG, Fay RR (2005) Development of the acoustic startle response of zebrafish to tones. J Exp Biol 208:1363–1372
Zeddies DG, Fay RR, Alderks PW, Shaub KS, Sisneros JA (2010) Sound source localization by the plainfin midshipman fish, *Porichthys notatus*. J Acoust Soc Am 127: 3104–3113
Zeddies DG, Fay RR, Gray MD et al (2012) Local acoustic particle motion guides sound-source localization behavior in the plainfin midshipman fish, *Porichthys notatus*. J Exp Biol 215:152–160

MEMBERS OF THE ASA COMMITTEE ON STANDARDS (ASACOS)

P.D. Schomer, ***Chair and ASA Standards Director***

Schomer and Associates
2117 Robert Drive
Champaign, IL 61821
Tel: +1 217 359 6602
Fax: +1 217 359 3303

R.D. Hellweg, Jr., ***Vice Chair***

Hellweg Acoustics
13 Pine Tree Road
Wellesley, MA 02482
Tel: +1 781 431 9176

S.B. Blaeser, ***Standards Manager***

Standards Secretariat
Acoustical Society of America
1305 Walt Whitman Rd., Ste. 300
Melville, NY 11747
Tel: +1 631 390 0215
Fax: +1 631-923-2875
Email: asastds@aip.org

Representation S1, Acoustics

R.J. Peppin, *Chair, S1*
A. Scharine, *Vice Chair, S1*
R.J. Peppin, *ASA Representative, S1*
R.D. Hellweg, *ASA Alternate Representative, S1*

Representation S2, Mechanical Vibration and Shock

C.F. Gaumond, *Chair, S2 ASA Representative, S2*
J.T. Nelson, *Vice Chair, S2 ASA Alternate Representative, S2*

Representation S3, Bioacoustics

C.J. Struck, *Chair, S3 ASA Representative, S3*
P. Nelson, *Vice Chair, S3 ASA Alternate Representative, S3*

Representation S3/SC1, Animal Bioacoustics

D.K. Delaney, *Chair, S3/SC1*
D.S. Houser, *Vice Chair, S3/SC1*
A. Thode, *ASA Representative, S3/SC1*

Representation S12, Noise

W.J. Murphy, Chair, S12
S.J. Lind, *Vice Chair, S12*
R.D. Hellweg, Jr. *ASA Representative, S12*
D. Lubman, *ASA Alternate Representative, S12*

ASA Technical Committee Representation

G.R. Potty, *Acoustical Oceanography*
M.M. Holt, *Animal Bioacoustics*
A.J. Campanella, *Architectural Acoustics*
P.J. Kaczkowski *and* W.E. Kreider, *Biomedical Acoustics*
R.M. Drake, *Engineering Acoustics*
A.C. Morrison, *Musical Acoustics*
R.J. Peppin, *Noise*
J. Mobley, *Physical Acoustics*
W.S. Woods, *Psychological and Physiological Acoustics*
C.F. Gaumond, *Signal Processing in Acoustics*
Z. Zhang, *Speech Communication*
R. Koch, *Structural Acoustics and Vibration*
R.M. Drake, *Underwater Acoustics*

Ex *Officio* Members of ASACOS

B.G. Shin-Cunningham, *Chair, ASA Technical Council* [= ASA Vice President]
D. Feit, *ASA Treasurer*
T.F.W. Embleton, *Past Chair ASACOS*
S.E. Fox, *ASA Executive Director*

U. S. Technical Advisory Group (TAG) Chairs for International Technical Committees

P.D. Schomer, *Chair, U. S. TAG, ISO/TC 43*
P.J. Battenberg, *U.S. Technical Advisor, U. S. TAG, IEC/TC 29*
W.M. Madigosky, *Chair, U. S. TAG, ISO/TC 108*

ANSI/ASA S3/SC1.4 TR-2014

MIX
Papier aus verantwortungsvollen Quellen
Paper from responsible sources
FSC® C105338
FSC
www.fsc.org

If you have any concerns about our products,
you can contact us on
ProductSafety@springernature.com

In case Publisher is established outside the EU,
the EU authorized representative is:
Springer Nature Customer Service Center GmbH
Europaplatz 3, 69115 Heidelberg, Germany

Printed by Libri Plureos GmbH
in Hamburg, Germany